THE WATER–FOOD–ENERGY NEXUS

The world of development thinkers and practitioners is abuzz with a new lexicon: the idea of "the nexus" between water, food, and energy which is intuitively compelling. It promises better integration of multiple sectoral elements, a better transition to greener economies, and sustainable development. However, there appears to be little agreement on its precise meaning, whether it only complements existing environmental governance approaches or how it can be enhanced in national contexts. One current approach to the nexus treats it as a risk and security matter while another treats it within economic rationality addressing externalities across sector. A third perspective acknowledges it as a fundamentally political process requiring negotiation amongst different actors with distinct perceptions, interests, and practices. This perspective highlights the fact that technical solutions for improving coherence within the nexus may have unintended and negative impacts in other policy areas, such as poverty alleviation and education.

The Water–Food–Energy Nexus: Power, Politics, and Justice lays out the managerial-technical definitions of the nexus and challenges these conceptions by bringing to the forefront the politics of the nexus, around two key dimensions – a dynamic understanding of water–food–energy systems, and a normative positioning around nexus debates, in particular around social justice. The authors argue that a shift in nexus governance is required towards approaches where limits to control are acknowledged, and more reflexive/plural strategies adopted.

This book will be of interest to academic researchers, policy makers, and practitioners in the fields of international development studies, environmental politics, and science and technology studies, as well as international relations.

Jeremy Allouche is a Research Fellow at the Institute of Development Studies at the University of Sussex, UK, and a member of the ESRC-funded STEPS Centre.

Carl Middleton is Director of the Center of Excellence in Resource Politics for Social Development in the Center for Social Development Studies (CSDS) at the Faculty of Political Science, Chulalongkorn University, Thailand.

Dipak Gyawali is Pragya (Academician) of the Nepal Academy of Science and Technology and was Nepal's Minister of Water Resources in 2002/2003. He conducts interdisciplinary research at the interface of technology and society, mostly from the perspectives of Cultural Theory.

PATHWAYS TO SUSTAINABILITY

This book series addresses core challenges around linking science and technology and environmental sustainability with poverty reduction and social justice. It is based on the work of the Social, Technological and Environmental Pathways to Sustainability (STEPS) Centre, a major investment of the UK Economic and Social Research Council (ESRC). The STEPS Centre brings together researchers at the Institute of Development Studies (IDS) and SPRU (Science Policy Research Unit) at the University of Sussex with a set of partner institutions in Africa, Asia and Latin America.

Series Editors:
Ian Scoones and Andy Stirling
STEPS Centre at the University of Sussex

Editorial Advisory Board:
Steve Bass, Wiebe E. Bijker, Victor Galaz, Wenzel Geissler, Katherine Home-wood, Sheila Jasanoff, Melissa Leach, Colin McInnes, Suman Sahai, Andrew Scott

One Health
Science, politics and zoonotic disease in Africa
Edited by Kevin Bardosh

Grassroots Innovation Movements
Adrian Smith, Mariano Fressoli, Dinesh Abrol, Elisa Around and Adrian Ely

Agronomy for Development
The Politics of Knowledge in Agricultural Research
James Sumberg

Epidemics
Science, governance and social justice
Edited by Sarah Dry and Melissa Leach

The Water-Food-Energy Nexus
Power, Politics, and Justice
Jeremy Allouche, Carl Middleton, and Dipak Gyawali

The Circular Economy and the Global South
Sustainable Lifestyles and Green Industrial Development
Edited by Patrick Schröder, Manisha Anantharaman, Kartika Anggraeni and Tim Foxon

"Beyond the commonplace recognition that the 'nexus' conceptual basis is not new and that integrative imperatives already featured in IWRM, this book further examines the underbelly of the beast and convincingly exposes the political underpinnings of a concept presented as a-political and 'manageable' through integrative tools, expert modeling, bureaucratic reforms, and rational efficiency-driven thinking. It reveals the underlying business imperatives and green economy logics, traces the global diffusion of the concept, and emphasizes that issues of distributional justice, knowledge production, and democratization of governance need to take center stage if the concept is to be transformative rather than supporting the status quo. An excellent text for all students of water and scholars interested in deciphering the world of water concepts and the interests and values that undergird them."

— Francois Molle, Editor of *Water Alternatives*

"We frequently hear of the nexus – but what does this mean, what does it entail, and where to begin? To such questions, Allouche, Middleton and Gyawali offer a critical guide. Careful to consider complexities and uncertainties, the theoretical discussion coupled with multi-sited case studies offers a compelling treatment. Readers wanting to know more of the concept, including political economic and equity implications, will find reading the book to be time well spent."

— Leila M. Harris, University of British Columbia, Canada

"Skilfully delving into the nuances of the nexus approach, the authors trace and explain the emergence of the 'new' concepts of nexus – between water, food, energy, environment and more. Unravelling the tangle of nexus-invoking discourses, motivations and practices yields a valuable, sense-making analysis."

— John Dore, Australia's Department of Foreign Affairs and Trade

THE WATER–FOOD–ENERGY NEXUS

Power, Politics, and Justice

Jeremy Allouche, Carl Middleton, and Dipak Gyawali

Routledge
Taylor & Francis Group

LONDON AND NEW YORK

First published 2019
by Routledge
2 Park Square, Milton Park, Abingdon, Oxon OX14 4RN

and by Routledge
52 Vanderbilt Avenue, New York, NY 10017

Routledge is an imprint of the Taylor & Francis Group, an informa business

British Library Cataloguing in Publication Data
A catalogue record for this book is available from the British Library

Library of Congress Cataloging-in-Publication Data
Names: Allouche, Jeremy, author. | Middleton, Carl, author. |
Gyawali, Dipak, author.
Title: The water-food-energy nexus : power, politics and justice / Jeremy
Allouche, Carl Middleton and Dipak Gyawali.
Description: Abingdon, Oxon ; New York, NY : Routledge, 2019. |
Series: Pathways to sustainability series | Includes bibliographical
references and index.
Identifiers: LCCN 2018052566 (print) | LCCN 2019001766 (ebook) |
ISBN 9781315209067 (eBook) | ISBN 9781138714274 (hbk) |
ISBN 9780415332835 (pbk) | ISBN 9781315209067 (ebk)
Subjects: LCSH: Sustainable development. | Water security. |
Food security. | Energy security.
Classification: LCC HC79.E5 (ebook) |
LCC HC79.E5 A434 2019 (print) | DDC 338.9--dc23
LC record available at https://lccn.loc.gov/2018052566

ISBN: 978-1-138-71427-4 (hbk)
ISBN: 978-0-415-33283-5 (pbk)
ISBN: 978-1-315-20906-7 (ebk)

Typeset in Bembo
by Taylor & Francis Books

CONTENTS

ILLUSTRATIONS

Figures

Tables

ACKNOWLEDGEMENTS

The authors would collectively like to thank the project "Dams, securitization, risks and the global water–energy nexus under climate change scenarios (KN/11015)", of the Social, Technological and Environmental Pathways to Sustainability (STEPS) Centre, Institute of Development Studies, University of Sussex for its support in preparing this book, and in particular Melissa Leach, Ian Scoones, and Andy Stirling. We would also like to thank Lina Forgeaux for copy-editing. We are also grateful to the Routledge editorial team, in particular Leila Walker, for her patience and professional guidance. Middleton would like to thanks his colleagues at the Center for Social Development Studies (CSDS) of the Faculty of Political Science, Chulalongkorn University including Ake Tangsupvattana, Naruemon Thabchumpon, and Orapan Pratomlek, together with Kanokwan Manorom and Surasom Krisanajhutha of the Mekong Sub-Region Social Research Centre (MSSRC) of the Faculty of Liberal Arts, Ubon Ratchathani University, for their support over the project, including co-organising the "The Nexus in Asia" panel discussions in September 2014. Gyawali wants to thank the main driving spirits behind the 2015 February Fulbright Water–Energy–Food Nexus Workshop in Kathmandu: Laurie Vasily, then director of the US Educational Foundation which ran the Fulbright program, Ari Nathan of the US Embassy regional environmental office for South Asia based in Kathmandu, and Philippus Wester of the International Center for Integrated Mountain Development (ICMOD). It helped greatly in sharpening some of the arguments and concepts behind the nexus as applied in South Asia.

1

INTRODUCTION

Nexus and nexuses

The world of development thinkers and practitioners is abuzz with a new lexicon: the idea of "the nexus" between water, food, and energy. In Asia and globally, the water–food–energy nexus has received growing attention from policy makers, researchers, and practitioners. The Scopus database recorded more than 221 peer-reviewed English language articles on the nexus in 2016 (Albrecht, Crootof, & Scott 2018). The nexus concept is thus the latest development paradigm shaping the world of resource management. A key premise of the nexus is that water use is interdependent with energy and food production. It proposes that these systems are inextricability linked, and thus integrated approaches are required that move beyond sectoral, policy, and disciplinary silos. Thus, an idea that started at the World Economic Forum (WEF) in 2008 has gained salience over time, through the Bonn Conference in 2011, the Sixth World Water Forum in Marseille, France, the Rio+20 negotiations in 2012, and the 2014 Stockholm Water Week to become the new vocabulary defining sustainable development amongst this sphere of policy makers. Unlike in the past, with environmental activists driving the discourse, this time it was also corporate actors that have played a key role in defining dominant debates around the nexus, resource efficiency, and scarcity.

Governing the nexus is probably one of the grand challenges of the 21st century. Who could deny that the nexus of water, food, energy, and the environment somehow encapsulates some of the world's most pressing problems and that governance is a key part of the problem as well as the solution? Is there something to disagree about when the international business community, through the WEF, argues that there are important linkages between water, food, energy, and climate change? And when the German government argues that policy makers need to consider more carefully the trade-offs between these four different resources? Isn't the idea of the nexus, after all, intuitively compelling, even as it challenges existing comfortable knowledge and approaches that have hitherto guided institutions managing these resources in independent silos?

In this book, we will argue that the nexus is not a straightforward policy solution, in that there are many ways in approaching these sectoral relationships. It is therefore more appropriate to talk about nexuses in plural. In this introduction, we will see how the idea of the nexus reflects a political process rather than a technical solution, which represents a particular framing of the problem of resource management at a critical juncture. It is important to recall perhaps more fundamentally that nexus thinking, or integrated approaches, are not new.

1.1. Nexus: A political process

Water, energy, climate, and food security, which are highly important for society, are evidently closely related – but what exactly is meant by the nexus? While the nexus is often presented as the integration of multiple sectoral elements such as energy, climate, water, and food production within an over-arching governance approach (Weitz et al 2017), there appears to be little agreement on its precise meaning. Elements of the nexus concept as an ideal include aspirations for understanding and managing scarcity, synergies, and trade-offs; increasing efficiency; bridging across fragmented food, water, and energy policy and institutions; improving governance; and ultimately ensuring that development is sustainable. Whilst each of these concepts is broadly appealing, the idea of the nexus has traits of a "nirvana concept" (Molle 2008), analogous to the idea of Integrated Water Resources Management (IWRM). As Molle (2008: 131) states:

> [i]deas are never neutral and reflect the particular societal settings in which they emerge, the worldviews and interests of those who have the power to set the terms of the debate, to legitimate particular options and discard others, and to include or exclude particular social groups.

Leach, Scoones, and Stirling (2010: 43–52) and others (see, for example, Molle 2008; Walker 2012: 4–5) highlight that there are many ways of explaining a socio-technical-environment system with equally rational ways of understanding it. This, in turn, can lead to different narratives of explanation between actors of the same system. Narratives are causal and explanatory beliefs (Molle 2008) produced by actors that frame systems in particular ways towards attaining specific goals. The construction of frames involves subjective normative judgements and choice of elements. Thus, framing recognises that any system is subject to multiple forms of interpretation by a range of actors dependent upon how scale, boundaries, key elements, dynamics, and outcomes are labelled and categorised, and how assumptions are made based on varying degrees of subjective/value judgements. Molle (2008) shows how the ideational power of nirvana concepts underpins the construction and framing of narratives.

Recognition of the nexused relationship between food, water, and energy has the potential to add significant value towards resource-management policy and practice. The point is that there is a need to acknowledge the existence and legitimacy of a range of narratives and frames in pursuing a nexus approach given the complexity and diversity of relationships within any nexused water–food–energy system, and the number of actors with divergent interests involved; in other words, the nexus is a political process, one where the interplay of different types of power as well as actors wielding them, is not just a procedurally technical one. Nexus thinking, in the form of integrating water security with agriculture, energy production, and climate concerns, is normatively argued to provide a means to better transition societies towards greener economies and the wider goal of sustainable development. Green economy aims to stimulate green investments, revive economic growth, and decouple growth from environmental decline, especially from fossil fuel addiction. In most green-economy narratives and in conservationist accounts of the Anthropocene, nature is portrayed as scarce, constrained by "planetary boundaries", and under siege by an undifferentiated "humanity". Green economy aims to elevate science over conventionally entrenched politics, although this rendering technical itself is a highly political move. To this end, nature's components are to be dis-aggregated and quantified; regulated and rented out by states; and in some versions, privatised, priced, and traded transnationally (e.g. in carbon and biodiversity markets). Advocates see this as efficient strategy in a time of climate emergency and as a way to surmount North–South conflicts and intractable issues of inequality and resource control. Yet, the politics of aspirations for "green transformations" in society – as aspired to under the "green economy" – are increasingly the subject of critical research agendas (e.g. Scoones, Leach, & Newell 2015). Though the idea of "greening" intimately and intuitively links with multiple dimensions of sustainability, it, however, almost inevitably raises questions with respect to socio-ecological processes of inequalities and access to livelihoods, paying attention to both distribution of resources and an examination of benefits and burdens of these large-scale "green" projects. There are numerous studies in political ecology that have already highlighted how interventions narrowly aimed at environmental conservation can lead to exclusion and dispossession for people, communities, and states, undermining livelihoods and creating and intensifying economic, social, political, and sometimes gender inequities (e.g. Robbins 2011). With regard to the nexus concept, several key issues emerge from the current debate including the extent to which such conceptualisations are genuinely novel, whether they complement (or are replacing) existing environmental governance approaches, and how – if deemed normatively desirable, including from a social justice perspective – the nexus can be enhanced in local, national and transnational contexts.

1.2. Nexus: Nothing new so why now?

In its most recent incarnation since the WEF in 2008, the nexus has been put forward as a new framing of the interdependent problems of water, food, and energy, demanding innovative solutions. However, there is nothing new; the modellers, the farmers, or the civil engineer have known about these relationships for a long time. Furthermore, the idea of the nexus has been talked about before in academic circles, and in particular amongst those researching water resources; Gleick (1994), for example, wrote about the nexus between energy and freshwater, arguing that there are strong connections between the growing water crises and conflicts over energy resources and that they should be considered together for policy and decision-making. As a policy challenge, nexus terminology appears to have begun in 1983 with the Food–Energy Nexus Programme of the United Nations University (UNU), which sought to better understand the relationships between food and energy challenges in developing countries and the identification of technical and policy solutions (Sachs & Silk 1990). They were also two major international conferences on the topic of nexus during the same period, that were relating the nexus to ecosystems challenges, namely the Food, Energy, and Ecosystems conference, which was held in Brasilia, Brazil in 1984 (Alam 1988) and the Second International Symposium on the Food–Energy Nexus and Ecosystems, which was held in New Delhi, India, February 12–14, 1986 (Parikh 1986). The first conference was held under the joint auspices of the UNU, United Nations Educational, Scientific and Cultural Organization (UNESCO), and the following Brazilian institutions: FINEP (Brazilian Agency for Research Financing), CNPq (Conselho Nacional de Desenvolvimento Científico e Tecnológico, The Brazilian National Council for Scientific and Technological Development), EMBRAPA (Brazilian Agency for Agricultural Research), and the University of Brasilia. It was attended by more than 80 researchers from throughout Latin America and leading experts from other continents. The principal outcome of the conference was the identification of inputs for the elaboration, evaluation, and installation of integrated food–energy projects as well as the development of appropriate research programmes. In Brazil, this conference led to subsequent research and teaching programmes on integrated food–energy systems (Sachs & Silk 1990). A major follow-up to the 1984 Brasilia conference was the Food–Energy Nexus and Ecosystem conference held in February 1986 in new Delhi. It was organised by the Indian Institute of Management (IIM) with the additional support of the Indian Department of Non-conventional Energy Sources (DNES) and UNESCO. It compared the design and operation of existing integrated food–energy production systems developed in different countries under diverse ecological and socio-economic conditions. After the New Delhi conference, Drs. T.K. Moulik (IIM, India) and Emilio La Rovere (FINEP, Brazil) jointly prepared in June 1987 a feasibility report for "Establishing a Permanent International Network on Biomass-based Agro-Industrial-Energy Systems" designed to advance South–South co-operation on integrated food energy systems.

So why does suddenly the nexus appear as a key policy solution? Nexus thinking emerged at a critical moment against the backdrop of cascading global crises in energy, food, and global finance in 2008, and a sense of scarcity and insecurity that accompanied it, especially amongst transnational corporations and the Organisation for Economic Co-operation and Development (OECD) country governments. For instance, Nestlé has been quite vocal about the issue. In a leaked report on a meeting organised with the US Embassy in Switzerland, Herbert Oberhaensli, Nestle's chief economist and director of international relations at the time, considers the water economy – and not the current financial crisis, oil depletion, or global warming – to be the most dangerous near-term threat to the planet's wellbeing. A similar point was again made by Nestlé chairman Peter Brabeck-Letmathe in an interview given to the *Financial Times*. In another interview with McKinsey, his tone becomes alarmist, declaring that "The water crisis that seems possible within the next 10 to 20 years will therefore quite probably trigger significant shortfalls in cereal production and, as a result, a massive global food crisis".[1] This is based on Nestlé's assumption that the earth's maximum sustainable fresh water withdrawals are about 12,500 cubic kilometres per year. In 2008, global fresh water withdrawals reached 6,000 cubic kilometres per year. All the available supply should be used by 2050, according to their scenarios.

These were further amplified by the "uncertainties" brought about by climate change, which were slowly becoming visible to the policymakers (IPCC 2014). The link between the nexus and climate change was already recognised at the 2008 WEF, where climate change risks for water/food/energy sectors could be worsened if contradictions in water and energy resource use are overlooked (Allan, Keulertz, & Woertz 2015). The coordination between the water, energy, and agriculture sectors was perceived as a way to promote effective climate change adaptation strategies, given that the impacts of and responses to climate change are generally cross-sectoral (England et al 2017; Mohtar & Lawford 2016; Pardoe et al 2018). The nexus is now becoming more central to climate change policy debates, as it is clear that through changes in water availability, climate change will influence both agriculture and energy production (OECD 2014). There is therefore an emerging literature on the conceptual and empirical relationship between the nexus and climate change (see Azhoni, Holman, & Jude 2017; Conway et al 2015). There has been a strong empirical focus on southern Africa, which is for some experts a region emblematic of the connections between climate and the water–food–energy nexus, especially as the result of the impact of climate change on food security (see Carter & Gulati 2014; Conway et al 2015). Other studies have focused on these interconnections in Central Asia (Maas et al 2012) or Sardinia (Masia et al 2017). Nexus or interlinked governance across sectors has been fronted as a potential solution to the spectre of scarcities in food and energy sectors, and social changes that included population growth, globalisation, economic growth, and urbanisation, as well as to potentially tackle the anomalies that climate change would bring in future (Hoff 2011).

While the surge in climate related uncertainties have, on the one hand, brought back fears of "scarce" resources and limits to growth, on the other hand, they have also exacerbated fears of an unforeseeable future which is difficult to predict and control as its impacts on natural and ecological cycles remain uncertain. However, these uncertainties are increasingly dealt with through "stability and durability thinking", encouraging the construction of large manmade structures rather than "resilience and robustness thinking", where the limits to control are acknowledged and adaptive solutions that incorporate plural solutions are pursued to address uncertainty and complexity. Drawing on Cultural Theory, these plural solutions might be considered as "clumsy solutions" that benefit from the deliberate interaction of multiple world-views on perceptions of problems and associated risk, ways of organising social rela-tions, and ultimately generating creative, new solutions and alternatives and plural policy responses (Gyawali 2009; Verweij et al 2006).

Overall, the nexus debate is primarily a debate about natural resource scarcity (Dupar & Oates 2012). The nexus in its present form is a response to the global crisis that hit energy and food demands in 2007 and 2008. In a paradoxical way, this was the first time that the business community, essentially through the WEF, came to realise the limits to growth. It is we will argue in this book, however, the existence of scarcity and its embedding as a concept in discourse is itself highly political and relates as much to socially-produced scarcity emerging from principles, processes, and politics of distribution, as it does to absolute scarcity (Mehta 2010). Scarcity has long been a totalising discourse in resource politics and mainstream economics (Mehta, Huff, & Allouche 2018). It is a technical constant given *a priori* that is rarely questioned. In academic and policy debates, the idea of scarcity has become uni-versalised and even naturalised, i.e. a powerful and taken-for-granted idea that scar-city is a natural phenomenon that exists outside of human society and politics, and can be isolated from planning models, allocation politics, policy choices, market forces, and local power, social, and gender dynamics (see Mehta 2010; Xenos 1989). The fact is that, besides physical scarcity occurring in particular geographies and seasons, it is also socially constructed − as a means of bureaucratic control, or a means of market players selling particular solutions or even scaremongering to elicit alarmist behaviour engaged in by activist campaigns. It is when scarcity becomes an instru-mentalised and a hegemonic discourse that things become problematic. Particular forms of scientific knowledge, technology, governance, market mechanisms, and innovation are evoked as the appropriate (and often, only) solutions. Indeed, this is the central problem of economics, and an underlying justification for industrial capitalism and even the nation-state as a form of social and economic organisation and control (Conrad & Clark 1987; Stiglitz 1988; Xenos 1987). However, there are a multitude of unexamined assumptions about the nature of scarcity as a problem, about the technologies and ostensible fixes that are put forward as solutions and the implications of this trajectory for socio-environmental dynamics (see Xenos 1989).

1.3. Nexus or nexuses?

This growing scarcity narrative is fuelling a new policy paradigm trying to respond to these crises. But what does this policy solution entail? The nexus is still very much an immature, evolving concept, and there are multiples framings in compe-tition. However, we can see some core perspectives emerging, including on: risk and security, and economic rationality.

Within the literature, there are nexuses, not just one nexus, with divergent framings between various proponents (Bizikova et al 2013). Differences are apparent in the empirical foci of research and terminologies employed. By no means exhaustive, these include inter alia: the "water–food–energy–climate nexus" (Beck & Villarroel Walker 2013; Pittock, Hussey, & Dovers 2015; WEF 2011); the "water–food nexus" (Mu and Khan 2009); the "water–energy nexus" (Hussey & Pittock 2012; Kouangpalath & Meijer 2015; Perrone, Murphy, & Hornberger 2011; Scott et al 2011); the "energy–water nexus" (Marsh & Sharma 2007; Murphy & Allen 2011; Stillwell et al 2011); the "bioenergy and water nexus" (UNEP 2011); the "energy–irrigation nexus" (Shah et al 2003); "water–energy–food security nexus" (Bazilian et al 2011; Bizikova et al 2013; ICIMOD 2012; Lawford et al 2013); water–energy–food ecosystems nexus (De Strasser et al 2016); and related concepts such as "land use-climate change-energy nexus" (Dale, Efrovmson, & Kline 2011) and a range of development related nexus approaches (see Groenfeldt 2010). One can see that the actual number of nexus sectors also differs, focusing sometimes only on two sectors (water–energy or water–food) or extended to additional sectors such as climate change, ecosystems, or livelihoods. Besides the range of nexuses, one can identify various scales, ranging from cities (Endo et al 2015; Hoff 2011; Villarroel Walker et al 2014), to transboundary river basins (Bach et al 2012; De Strasser et al 2016; Kibaroglu & Gürsoy 2015; Lawford et al 2013; McLachlan 2015). The concept of the nexus – and its associated research methodologies – is therefore far from unified and seemingly varies according to the focus of sectoral integration studied. Some terminologies adopt an energy, climate, or food focus but all these sectors are invariably linked to water resource management.

The first core perspective of the nexus that we see emerging addresses risk and security. The dominant global policy-framing of the nexus emphasises a scarcity-crisis narrative producing a sense of growing water, food, and energy insecurities (Lebel & Lebel 2017). Resource scarcity due to the interconnectedness of these resources could lead to conflicts (see e.g. Bizikova et al 2013; Hoff 2011; WEF 2012), very much in line with the environmental scarcity-conflict logic of the early 1990s (Homer-Dixon 1994; on the limits of this logic, see Huff 2017). These water, food, and energy insecurities have been growing due to new projects, such as the production of first-generation biofuel and its impact on food security (Rulli et al 2016), or irrigation diversion and its impact on water security (Blake 2016). There is further research which argues that water–food–energy security and political stability are significantly correlated and that insecurities for water, food, and energy are independently destabilising and pose increased risk combined (Abbott et al 2017). Proposed strategies for reducing nexus-related security risks has been the promotion of large-scale control-type technologies of mass production for meeting food and energy demands in the belief that they are more secure. This approach, in practice, can result in the redistribution of access to natural resources away from small-scale users towards large bureaucracies or corporations.

The second perspective is guided by economic rationality. Hoff (2011) suggests that nexus thinking is concerned with addressing externalities across sectors. Instead of a focus on productivity within a sector, a nexus approach focuses on overall system efficiency. Similarly, Rasul (2016) describes the nexus approach as "a system-wise approach that recognises the inherent interdependencies of the food, water, and energy sectors for resource use". Here the nexus is seen as a way to improve policy cost-effectiveness and resource-use efficiency, as well as to optimize allocation of resources across sectors (SIWI 2014). Sustainable water use in energy and food supply chains, it is believed, will create new business opportunities around green economic growth (Vlotman & Ballard 2014; Wales 2014; Zahner 2014). In the nexus perspective here, policy coherence is undermined by the sectors' different institutional frameworks, divergent targets, lack of communication, and lack of clarity on rights and responsibilities across sectors (Pittock, Hussey, & McGlennon 2013). The proposed policy solution is primarily embedded in the logic of environmental economics, and its recent policy manifestation, green economy, which gives primacy to resource valuation, pricing and value chain optimisation (Death 2014). Policy solutions tend to draw upon language that frames natural resources as natural capital, leading to its monetisation. In the end, this nexus approach is characterised by its tendency towards decision-making tools that follow a system approach where the interactions between the different sectors are quantified and modelled as global and regional flows for their optimisation, although this approach sometimes ignores the day-to day realities and local priorities and needs in terms of human security. The problematic assumption in this understanding of the nexus is that water, energy, food needs are viewed as a trade-off that can be dealt with by a perfect equilibrium model from a socio-ecological system thinking perspective through which resource allocation can be decided. Different institutional frameworks in these hitherto differently treated sectors are seen as passive and amenable to easy re-engineering instead of being seen as active actors themselves who will resist any change imposed upon them that is not commensurate with their way of doing things. This depoliticised and decontextualised approach through aggregate modelling and forecasting techniques fails to engage with the local realities linked with these trade-offs. Furthermore, a challenge in this approach is that aiming for optimal use can stifle space for more innovative solutions to take place and in the process can create considerable risks. Yet, we would suggest, diverse responsive policy solutions are also needed to manage the nexus in a time of risk and uncertainty.

1.4. A different perspective on the nexus

There is an important issue which is ignored in these two framings of the nexus. The food, energy, and financial crisis reveals broader structural issues in the functioning of the current system. As pointed out by Leese and Meisch (2015), the proposed measures presented in various nexus reports reveal that the nexus – and the crisis that it is to address – is in fact conceived of as something that is very much manageable, even as globally many of the earth's ecological systems are degraded and under pressure (Hoff 2011: 4; WRG 2009: 6). As put by Foran

(2015), commentators were already in 2012 (Hoff 2011) referring to the need to "manage" the nexus as if it had become a relatively well-defined class of problems, one that could be quantified using integrated assessment tools, and addressed via capacity building and other improved managerial and governance responses. Yet, doesn't the crisis entail the need for transformations towards a different system (Leese & Meisch 2015), based on a rather different set of values that would be cognizant of the many environmental and social injustices present within the current situation? Indeed, it is likely that the nexus "crisis" narrative will further deepen existing inequalities in the current system rather than address them.

Furthermore, there is still little debate concerning who is making current decisions within a nexus framing, and in whose interest. We suggest that much of the current nexus framing, located in international business imperatives and global (neoliberal) policy, masks different types of politics:

- Politics of difference (inequality, class, gender, identity within and between societies)
- Politics of knowledge (multiple framings, and positionalities, subjectivities, the manufacture of scarcities)
- International political economy and geopolitics (resource grabs, whose limits/ boundaries count, North–South geopolitics)

Presently, governing the nexus is interpreted in much of the literature as principally a technical or administrative matter, where better coordination of information about cross-sector interactions can improve, or even optimise, system performance as measured against security or economic criteria. However, information alone does not necessarily lead to policy change and administrative processes are not necessarily objective (Kurian 2017). In this book, we propose a counter position that offers an alternative perspective on the nexus. Contributing to an emerging literature, we make the case that addressing trade-offs and improving policy integration across sectors is a fundamentally political process requiring negotiation amongst different actors with distinct perceptions, interests, and practices (Allouche, Middleton, & Gyawali 2014; Rees 2013; Stein, Barron, & Moss 2014). It calls into question these apolitical technical solutions by highlighting their impacts in other policy areas, most notably poverty alleviation (Jobbins et al 2015). This perspective takes a normative stance, around equity and social justice (Dupar & Oates 2012; Stringer et al 2014). Nexus thinking should be used to highlight linkages between local resources, livelihoods, and rights and social justice, not just between sectors such as food, water, energy (see for example Harper-Dorton & Harper 2015) and not just with the comfortable tools available within them. Power, politics, and justice, as reflected in the title of this book, are the key barriers and entry points for governing the nexus that demand the generation of new and uncomfortable knowledge as well as tools more amenable to analysing political trade-offs.

There is a need for critical social science analysis and conceptualisation of the nexus. This includes pursuing a political economy/political ecology perspective that explicitly addresses power relations, and normative aspects of the nexus, in particular social justice. This book therefore lays out the managerial-technical definitions of the nexus and challenges its conceptions by bringing to the forefront the politics of the nexus around two key dimensions – a dynamic understanding of water–food–energy systems, and a normative positioning around nexus debates. Water–food–energy–society systems are inherently complex and dynamic, especially under the conditions of climate change. Yet, as we have outlined above, a lot of nexus literature tends to see the system's management as largely within human's ability to manage through direct-control measures that tends towards large-scale hard solutions, such as large infrastructure projects. We argue that a shift in nexus governance is required towards approaches where limits to control are acknowledged, and more reflexive/plural strategies adopted that draws on "non-equilibrium thinking" (Leach, Scoones, & Stirling 2010). We begin with the premise that uncertainty is an intrinsic element of nature itself (Penrod 2001), and that nexus solutions overlook – or at least significantly downplay – this element. Drawing on a range of local and cross-border case studies in Asia, we call for an alternative more-grounded framing of the nexus. Methodologically, we draw on a multi-sited analysis emphasising the political economy and justice dimensions of the nexus: globally, tracing the nexus discourse and its diffusion; regionally, understanding the political economies and imaginaries of 'the nexus' in South and South East Asia; and project wise, examining case studies that exemplify nexused-dilemmas and where divergent, plural perspectives and contestations have emerged in response.

The goal of this book is to explore the implications of the global nexus policy discourse, and how it is being translated into regional research and policy agendas. We unpack: who is defining and promoting the water–food–energy nexus, how and why?; What have been the intended and unintended outcomes to date?; Is the nexus displacing, complementing, or replacing the IWRM paradigm that emerged in the 1990s?; and, How can the concept of dynamic sustainability be sharpened by political economy/ecology and cultural theory thinking so as to better capitalise on and broaden the current nexus discourse? Overall, we argue that a different framing of the nexus is required: one which recognises that global priorities may not reflect local concerns, recognising the politics of scales; and that resource allocations are political decisions shaped and limited by a society's degree of concern with social justice, which need to be decided through more open and transparent decision-making. Thus, for the nexus to become more inclusive as a policy agenda, it must first be grounded in local realities and human needs, and far more concerned with social justice, which links it with concerns of ethics.

Note

1 www.mckinsey.com/business-functions/sustainability-and-resource-productivity/our-in sights/water-as-a-scarce-resource-an-interview-with-nestl-and-233s-chairman (accessed 10 August, 2018).

References

Abbott, M., Bazilian, M., Egel, D., & Willis, H. H. (2017). Examining the food–energy–water and conflict nexus. *Current Opinion in Chemical Engineering*, 18, 55–60.

Alam, A. (1988). Energy requirements of food production and utilization in the rural sector. In Food-energy nexus and ecosystem: Proceedings of the Second International Symposium on Food-Energy Nexus and Ecosystem, held in New Delhi, India, during February 12–14, 1986 (vol. 131, p. 270). Columbia, MMO: South Asia Books.

Albrecht, T. R., Crootof, A., & Scott, C. A. (2018). The water-energy-food-nexus: A systematic review of methods for nexus assessment. *Environ. Res. Lett*, 13, in press.

Allan, T., Keulertz, M., & Woertz, E. (2015). The water-food-energy nexus: An introduction to nexus concepts and some conceptual and operational problems. *International Journal of Water Resources Development*, 31(3), 301–311.

Allouche, J., Middleton, C., & Gyawali, D. (2014). Nexus nirvana or nexus nullity? A dynamic approach to security and sustainability in the water-energy-food nexus, *STEPS Working Paper 63*. Brighton, UK: STEPS Centre, Institute of Development Studies.

Azhoni, A., Holman, I., & Jude, S. (2017). Adapting water management to climate change: Institutional involvement, inter-institutional networks and barriers in India. *Global Environmental Change*, 44, 144–157.

Bach, H., Bird, J., Jonch Clausen, T., Morck Jensen, K., Baadsgarde Lange, R., Taylor, R., Viriyasakultorn, V., & Wolf, A. (2012). *Transboundary river basin management: Addressing water, energy and food security*. Vientiane, Laos: Mekong River Commission.

Bazilian, M., Rogner, H., Howells, M., Hermann, S., Arent, D., Gielen, D., & Yumkella, K. K. (2011) Considering the energy, water and food nexus: towards an integrated modelling approach, *Energy Policy*, 39(12), 7896–7906.

Beck, M. B. and Villarroel Walker, R. (2013). On water security, sustainability, and the water-food-energy-climate nexus. *Frontiers of Environmental Science & Engineering*, 7(5), 626–639.

Bizikova, L., Roy, D., Swanson, D., Venema, H. D., & McCandless, M. (2013). *The water-energy-food security nexus: Towards a practical planning and decision-support framework for landscape investment and risk management*. Manitoba, Canada: The International Institute for Sustainable Development.

Blake, D. (2016). Iron triangles, rectangles or golden pentagons? Understanding power relations in irrigation development paradigms of Northeast Thailand and Northern Cambodia. *Water Governance Dynamics in the Mekong Region*, 23–58.

Carter, S. & Gulati, M. (2014). Climate change, the Food Energy Water Nexus and food security in South Africa. In *Understanding the Food Energy Water Nexus*. WWF-SA, South Africa.

Conrad, J. M. & Clark, C. W. (1987). *Natural resource economics: Notes and problems*. Cambridge, UK: Cambridge University Press.

Conway, D., Archer van Garderen, E., Deryng, D., Dorling, S., Krueger, T., Landman, W., & Dalin, C. (2015). Climate and Southern Africa's water-energy-food nexus. *Nature Climate Change*, 5, 837–846.

Dale, V. H., Efroymson, R. A., & Kline, K. L. (2011). The land use–climate change–energy nexus. *Landscape Ecology*, 26(6), 755–773.

Death, C. (2014). The green economy in South Africa: Global discourses and local politics. *Politikon*, 41(1), 1–22.

De Strasser, L., Lipponen, A., Howells, M., Stec, S., & Bréthaut, C. (2016). A methodology to assess the water energy food ecosystems nexus in transboundary river basins. *Water*, 8, 59.

Dupar, M., & Oates, N. (2012) *Getting to grips with the water-energy-food "nexus"*. London: Climate and Development Knowledge Network. Available online: http://cdkn.org/2012/04/getting-to-grips-with-thewater-energy-food-nexus/ (accessed on 10 July 2015).

Endo, A., Burnett, K., Orencio, P., Kumazawa, T., Wada, C., Ishii, A., Tsurita, I., & Taniguchi, M. (2015). Methods of the water-energy-food nexus. *Water*, 7, 5806–5830.

England, M. I., Dougill, A. J., Stringer, L. C., Vincent, K. E., Pardoe, J., Kalaba, F. K., & Namaganda, E. (2017). Climate change adaptation and cross-sectoral policy coherence in southern Africa (Sustainability Research Institute Paper No. 108). SRI Papers, University of Leeds. Retrieved from www.see.leeds.ac.uk/fileadmin/Documents/research/sri/work ingpapers/SRIPs-108.pdf.

Foran, T. (2015). Node and regime: Interdisciplinary analysis of water-energy-food nexus in the Mekong region. *Water Alternatives*, 8(1), 655–674.

Gleick, P. H. (1994). Water and energy. *Annual Review of Energy and the Environment*, 267–299.

Groenfeldt, D. (2010) Viewpoint – The next nexus? Environmental ethics, water policies, and climate change. *Water Alternatives*, 3(3), 575–586.

Gyawali, D. (2009). Pluralized water policy terrain: Sustainability and integration. *South Asian Water Studies Journal*, 1(2), 193–199.

Harper-Dorton, K. V., & Harper, S. J. (2015). Social and environmental justice and the water-energy nexus: A quest in progress for rural people. *Contemporary Rural Social Work*, 7(1), 12–25.

Hoff, H. (2011). Understanding the nexus. Background Paper for the Bonn 2011 Conference: The Water, Energy and Food Security Nexus. Stockholm: Stockholm Environment Institute.

Homer-Dixon, T. F. (1994). Environmental scarcities and violent conflict: Evidence from cases. *International Security*, 19(1), 5–40.

Huff, A. (2017). Black sands, green plans and vernacular (in) securities in the contested margins of south-western Madagascar. *Peacebuilding*, 1–17, 153–169.

Hussey, K. & Pittock, J. (2012). The energy–water nexus: Managing the links between energy and water for a sustainable future. *Ecology and Society*, 17(1), 31.

ICIMOD (International Centre for Integrated Mountain Development) (2012). *Contribution of Himalayan ecosystems to water, energy, and food security in South Asia: A nexus approach.* Kathmandu, Nepal: International Centre for Integrated Mountain Development.

Jobbins, G., Kalpakian, J., Chriyaa, A., Legrouri, A., & El Mzouri, E. H. (2015). To what end? Drip irrigation and the water–energy–food nexus in Morocco. *International Journal Of Water Resources Development*, 31(3), 393–406.

Kibaroglu, A. & Gürsoy, S. I. (2015). Water-energy-food nexus in a transboundary context: The Euphrates-Tigris river basin as a case study. *Water Int*, 40, 824–838.

Kouangpalath, P. & Meijer, K. (2015). Water-energy nexus in Shared River Basins: How hydropower shapes cooperation and coordination. *Chang. Adapt. Socio-Ecol. Syst*, 2, 85–87.

Kurian, M. (2017). The water-energy-food nexus: Trade-offs, thresholds and transdisciplinary approaches to sustainable development. *Environmental Science & Policy*, 68, 97–106.

Lawford, R., Bogardi, J., Marx, S., Jain, S., Pahl Wostl, C., Knüppe, K., Ringler, C., Lansigan, F., & Meza, F. (2013). Basin perspective on the water-energy-food security nexus. *Current Opinion in Environmental Sustainability*. 5, 607–616.

Leach, M., Scoones, I., & Stirling, A. (2010). *Dynamic sustainabilities: Technology, environment, and social justice.* Abingdon, UK: Earthscan.

Lebel, L. & Lebel, B. (2017). Nexus narratives and resource insecurities in the Mekong Region. *Environmental Science & Policy*, 90, 164–172.

Leese, M. & Meisch, S. (2015). Securitising sustainability? Questioning the "water, energy and food-security nexus". *Water Alternatives*, 8(1), 695–709.

Marsh, D. M. & Sharma, D. (2007). Energy-water nexus: An integrated modeling approach. *International Energy Journal*, 8(4), 235–242.

Maas, A., Issayeva, G., Rüttinger, L., & Umirbekov, A., with contributions from Raul Daussa (2012). *Climate change and the water-energy-agriculture nexus in Central Asia. Scenario Report.* Berlin: Adelphi.

Masia, S., Sušnik, J., Mereu, S., Spano, D., Marras, S., Blanco, M., & Trabucco, A. (2017). Water-Food-Energy nexus and climate change for multipurpose reservoirs in Sardinia, Dresden Nexus Conference 2017. Available from: www.researchgate.net/p ublication/313401527_Water-Food-Energy_nexus_and_climate_change_for_multipurp ose_reservoirs_in_Sardinia [accessed Sep 28 2018].

McLachlan, S. N. (2015). Implementing the water-energy-food nexus at various scales: Trans-boundary challenges and solutions. *Chang. Adapt. Socio-Ecol. Syst,* 2, 94–96.

Mehta, L., Huff, A. & Allouche, J. (2018). The new politics and geographies of scarcity. *GeoForum,* in press.

Mehta, L. (ed) (2010). *The Limits to scarcity: Contesting the politics of allocation.* London and Washington, DC: Earthscan.

Molle, F. (2008). Nirvana concepts, storylines and policy models: Insights from the water sector. *Water Alternatives,* 1(1): 131–156.

Mohtar, R. H. & Lawford, R. (2016). Present and future of the water-energy-food nexus and the role of the community of practice. *Journal of Environmental Study Science,* 6, 192–199.

Mu, J. & Khan, S. (2009). The effect of climate change on the water and food nexus in China. *Food Security,* 1(4), 413–430.

Murphy, C. F. & Allen, D. T. (2011). Energy-water nexus for mass cultivation of algae. *Environmental Science and Technology,* 45(13), 5861–5868.

OECD (2014). New perspectives on the water-energy-food nexus: Forum background note. Paris: OECD.

Pardoe, J., Conway, D., Namaganda, E., Vincent, K., Dougill, A. J., & Kashaigili, J. J. (2018). Climate change and the water–energy–food nexus: Insights from policy and practice in Tanzania. *Climate Policy,* 18(7), 863–877.

Parikh, J. K. (1986). From farm gate to food plate: Energy in post-harvest food systems in south Asia. *Energy Policy,* 14(4), 363–372.

Penrod, J. (2001). Refinement of the concept of uncertainty. *Journal of Advanced Nursing,* 34(2), 238–245.

Perrone, D., Murphy, J., & Hornberger, G. M. (2011). Gaining perspective on the water-energy nexus at the community scale. *Environmental Science and Technology,* 45, 4228–4234.

Pittock, J., Hussey, K., & McGlennon, S. (2013). Australian climate, energy and water policies: Conflicts and synergies. *Australian Geographer,* 44(1), 3–22.

Pittock, J., Hussey, K., & Dovers, S. (2015). *Climate, energy and water: Managing trade-offs, seizing opportunities.* New York: Cambridge University Press.

Rasul, G. (2016). Managing the food, water, and energy nexus for achieving the Sustainable Development Goals in South Asia. *Environmental Development,* 18, 14–25.

Rees, J. (2013). Geography and the nexus: Presidential address and record of the royal geographical society (with IBG) AGM 2013. *The Geographical Journal,* 179(3), 279–282.

Robbins, P. (2011). *Political ecology: A critical introduction.* Chichester, UK: John Wiley & Sons.

Rulli, M. C., Bellomi, D., Cazzoli, A., De Carolis, G., & D'Odorico, P. (2016). The water-land-food nexus of first-generation biofuels. *Scientific Reports,* 6, 22521.

Sachs, I. & Silk, D. (1990). *Food and energy: Strategies for sustainable development.* Tokyo: United Nations University Press.

Scoones, I.; Leach, M. and Newell, P. (eds.) (2015) *The Politics of Green Transformations.* Abingdon: Earthscan.

Scott, C. A., Pierce, S. A., Pasqualetti, M. J., Jones, A. L., & Montz, B. E. (2011). Policy and institutional dimensions of the water-energy nexus. *Energy Policy*, 39, 6622–6630.

Shah, T., Scott, C., Kishore, A., & Sharma, A. (2003). *Energy-irrigation nexus in South Asia: Improving groundwater conservation and power sector viability.* Research Report 70. Colombo, Sri Lanka: International Water Management Institute.

SIWI (2014). *World Water Week in Stockholm 2014: Overarching conclusions.* Stockholm, Sweden: Stockholm International Water Institute.

Stein, C., Barron, J., & Moss, T. (2014). *Governance of the nexus: From buzz words to a strategic action perspective.* Thinkpiece Series. London: The Nexus Network.

Stiglitz, J. E. (1988). *Economics of the public sector.* New York: Norton and Company.

Stillwell, A. S., King, C. W., Webber, M. E., Duncan, I. J., & Hardberger, A. (2011). The energy-water nexus in Texas. *Ecology and Society*, 16(1), 2. www.ecologyandsociety.org/vol16/iss1/art2/.

Stringer, L. C., Quinn, C. H., Berman, R. J., Le, H. T. V., Msuya, F. E., Orchard, S. E., & Pezzuti, J. C. B. (2014). Combining nexus and resilience thinking in a novel framework to enable more equitable and just outcomes. Sustainability Research Institute Paper 73.

IPCC (2014). *Climate Change 2014: Synthesis Report.* Contribution of Working Groups I, II and III to the Fifth Assessment Report of the Intergovernmental Panel on Climate Change [Core Writing Team, R. K. Pachauri and L. A. Meyer (eds.)]. Switzerland: IPCC, Geneva.

UNEP (United Nations Environment Programme) (2011). *The bioenergy and water nexus.* Nairobi: UNEP.

Verweij, M., Douglas, M., Ellis, R., Engel, C., Hendriks, F., Lohmann, S., & Thompson, M. (2006). Clumsy solutions for a complex world: The case of climate change. *Public Administration*, 84(4), 817–843.

Villarroel Walker, R., Beck, M. B., Hall, J. W., Dawson, R. J., & Heidrich, O. (2014). The energy-water-food nexus: Strategic analysis of technologies for transforming the urban metabolism. *J. Environ. Manag*, 141, 104–115.

Vlotman, W. F. & Ballard, C. (2014). Water, food and energy supply chains for a green economy. *Irrigation and Drainage*, 63(2), 232–240.

Wales, A. (2014). Making sustainable beer. *Nature Climate Change*, 4(5), 316.

Walker, G. (2012). *Environmental justice: Concepts, evidence and politics.* London and New York: Routledge.

Weitz, N., Strambo, C., Kemp-Benedict, E., & Nilsson, M. (2017). Closing the governance gaps in the water-energy-food nexus: Insights from integrative governance. *Global Environmental Change*, 45, 165–173.

WEF (2012). *Water security: the water-food-energy-climate nexus.* Washington, DC: Island Press.

WEF (2011). *Global risks 2011* (Sixth Edition). Cologny/Geneva: WEF.

WRG (2009). *Charting our water future: Economic frameworks to inform decision-making.* Water Resources Group 2030 www.mckinsey.com/App_Media/Reports/Water/Charting_Our_Water_Future_Full_Report_001.pdf (accessed on 20 May 2014).

Xenos, N. (1987). Liberalism and the postulate of scarcity. *Political Theory*, 15(2), 225–243.

Xenos, N. (1989). *Scarcity and modernity.* London: Routledge.

Zahner, A. (2014). *Making the case: How agrifood firms are building new business cases in the water–energy–food nexus.* REEP and FAO. Available at www. reep. org/making-casehow-agrifood-firms-a re-building-new-business-cases-water-energy-food-nexus (accessed 20 May 2017).

2

A CRITIQUE OF THE GLOBAL HEGEMONIC NEXUS NARRATIVES

The problem with global crises is that the dominant hegemonic thinking tends to gravitative towards an urgent and singular solution, compatible with the hegemonic cosmology and ignoring or downplaying other possibilities. This is what happened in the case of "the nexus" in 2008, when the World Economic Forum (WEF), pushed by major multinational businesses,[1] proposed "the nexus" based on a particular scarcity narrative. The WEF's nexus approach, later reproduced and diffused by other global and regional actors, strongly emphasised demand-led technological and market solutions that downplayed or ignored supply-side limits together with the political dimensions that shape control over and access to resources. The nexus discourse and language diffused globally, including in South and South East Asia, although it is competing with other water–food–energy related concepts such as Integrated Water Resources Management (IWRM) (Allouche 2016).

Water–food–energy systems are by nature complex and dynamic systems (Lindberg & Leflaive 2015; see Table 2.1), and even more so under the conditions of climate change. The current global policy framing of the nexus and this politics of urgency is driving "stability-" and "durability-type" solutions (Leach, Scoones, & Stirling 2010) that downplay or ignore more dynamic pathways to sustainability. By this, we mean that underlying proposed solutions around the nexus is the paradigm of control that is founded on a belief that humans can predict and manage (often through large infrastructure) the environment, and implicitly societies' relationship with it.

This chapter is divided into three sections. First, we will outline the limits to the scarcity narrative which seems to be currently driving the nexus. Second, we discuss the concept of the nexus, primarily the market-technical framing of the WEF, which is embedded in business imperatives and logic of environmental economics and trace its global diffusion, taking the example of South East Asia. Finally, we conclude by arguing in favour of dynamic and resilient systems that could provide

TABLE 2.1 Gross list of nexus relations

Nexus relations	Water	Energy	Food
Water		Desalination requires energy	Water for sanitation competes with water for food
		Withdrawal of groundwater requires energy	
		Energy is needed for waste water treatment	
Energy	Water reservoirs for energy production		Bio energy crops compete for land with food crops
	Fracking (and other types of energy) requires water		
	Bio energy crops need water		
Food	Crops need water	Fertiliser and pesticide use energy	
	Food production may lead to water pollution	Farm mechanisation uses energy	
	Water is used in processing	Energy is used in food chain and transport	

Source: Reinhard, S. et al. (2017) *Water-food-energy nexus: A quick scan.* Wageningen, Wageningen Economic Research, Report 2017–096: 11

sustainable pathways for resource use. Here, we embrace plural systems that accommodate varieties of large- and small-scale solutions in varied climes where social justice is given due recognition.

2.1. A world in crisis: Deconstructing the current policy dynamics behind the nexus

The emergence of a nexus policy paradigm was fuelled by fears linked to the 2007 and 2008 food and energy price shock (Allouche 2011). After 20 years of low food commodity prices, the price shock of 2007/08 brought agriculture, food production, and food security sharply back into the limelight. The food commodity price crisis and the ensuing food riots raised fears around food scarcity and food security.

The causes of the rapid rise in food prices was disputed (Piesse & Thirtle 2009). The growing demand for food from rapidly developing countries (in particular China), the high price of oil, and the conversion of many crops to biofuel – all of which created pressures on the demand side – were highlighted by some analysts (Royal Society 2008). For others, weather-related poor harvests, flawed food and development policies, speculation in global financial markets, and the legacy of "food wars" were also important factors (see Messer 2009). At the same time, a global energy crisis occurred in 2007, when oil prices increased dramatically. This oil crisis was demand-led, unlike previous oil shocks which were caused by sudden interruptions in exports from the Middle East. The deteriorating value of the US Dollar vis-à-vis the Euro (although this trend has now reversed) and the significant growth of the Asian economies, especially China and India and the sheer size of their domestic markets and their energy consuming industries, created a strong demand for oil, which pushed up the energy price index (Hamilton 2009).

Both crises led to a policy of urgency and emergency. Rising prices for staple foods (i.e. maize, rice, wheat) and soybeans provoked riots in more than 20 countries (inter alia Mexico, Morocco, Indonesia, Uzbekistan, Yemen, Guinea, Burkina Faso, Mauritania, and Senegal) and non-violent demonstrations in at least 30 more (Benson et al 2008; FAO 2008; Von Grebmer et al 2008). This created a number of responses by policy makers around the various short- and long-term impacts of these crisis. References to global risks became commonplace at this time. Announcing a new report in 2011, the UN's Food and Agriculture Organization (FAO) claimed:

> Widespread degradation and deepening scarcity of land and water resources have placed a number of key food production systems around the globe at risk, posing a profound challenge to the task of feeding a world population expected to reach 9 billion people by 2050.
>
> *FAO 2011, para 1*

The issue even became securitised (Floyd 2010). At the 2008 WEF in Davos (Switzerland), the now former World Bank President Robert Zoellick argued that "increased food prices and their threat—not only to people but also to political stability—have made it a matter of urgency to draw the attention it needs".[2] Sir John Holmes, the previous UN Undersecretary General for Humanitarian Affairs, also echoed this argument. Stability of food systems rather than access to food became the focus of policy attention. As a result, the US National Intelligence Council highlighted in its 2030 Global Trends report highlighted the important geopolitical consequences – for conflict, national security, and global economy – if the nexus was not properly managed (Salam et al 2017: 16). This scarcity discourse enabled the nexus to become a matter of emergency, and, with it, its proponents – OECD governments and transnational corporations in particular – called for an appropriate response (see Leese & Meisch 2015).

Besides these material shocks that created turbulence in the global economy and had severe consequences for the poor in the global South and North (see for example Ruel et al 2009), the policy context and debates were marked by alarmist discourses towards the ongoing growth – and security of – the food and energy sectors (see for example Schlör et al 2018).

In terms of energy, ongoing strong demand from Asian economies, especially China and India, have long-lasting global implications: if global demand continues to rise then correspondingly energy production must also grow substantially each year, even as constraints imposed by social, environmental, and geopolitical realities become more pronounced. According to the projections of the US Department of Energy, world energy output, based on 2007 levels, must rise 29 per cent by 2025 to meet anticipated demand (quoted in Klare 2011). The International Energy Agency considers that global energy consumption is projected to grow by close to 50 percent by 2035 and 80 per cent by 2050 (IEA 2010). Furthermore, rising concerns about the long-term availability and price of not only oil, but also gas and uranium, only add to the fears and perceptions of a global energy crisis, even as shale gas (controversially) opens up the possibility of new sources for the fossil fuel industry (Howarth, Santoro, & Ingraffea 2011; Howarth, Ingraffea, & Engelder 2011).

In terms of food, the situation is expected to be exacerbated in the near future as 60 per cent more food will need to be produced in order to feed the world population in 2050. Total global water withdrawals for irrigation are projected to increase by 10 per cent by 2050 (FAO 2011). Furthermore, many scientific experts have argued that food management will be facing major challenges due to increasing uncertainties caused by climate change and its ecological consequences and by fast changing socio-economic conditions, including global redistributions of wealth and power, people migration, and shifting flows of resources and knowledge (Hanjra & Qureshi 2010; Schmidhuber & Tubiello 2007). According to the Intergovernmental Panel on Climate Change (IPCC) predictions, as climate extremes are anticipated to increase in frequency and intensity in the future, droughts and floods will become more severe and more frequent (IPPC 2007). This will have an effect on global food security as weather extremes can dramatically reduce crop yields and livestock numbers and productivity, especially in semi-arid areas. Most studies have found that climate change and other associated global environmental changes will have a highly negative impact for developing countries in terms of crop productivity; from 9 to 21 per cent by 2050 depending on the degree of change modelled (Misselhorn et al 2012). This raises important concerns about achieving food security, especially for politically and economically marginalised people. Some of the poorest regions with high levels of population growth as well as of chronic undernourishment will also be exposed to the highest degree of instability in food production (Allouche 2011). Climate change may affect food systems in several ways ranging from direct effects on crop production as discussed above to changes in markets volatility, food prices, and supply chain infrastructure (Schmidhuber & Tubiello 2007). Furthermore, the nature and timing of climate impacts on agriculture and their implications for human livelihoods are clouded with many uncertainties (Hertel, Burke, & Lobell 2010). A key concept and policy response to these scenarios has been that of "sustainable intensification", which is about increasing food production using the same area of land, while minimising pressure on the environment (Rockström et al 2017). However, there are a number of concerns that this concept might be used to justify intensification *per se* via the accelerated adoption of high-input and hi-tech agriculture (see for example Godfray 2015).

The nexus discourse has been evidenced from the research and narratives produced from these crisis and scenarios, due to their intertwined relationship. For example, an estimated 30 to 50 per cent of the food produced globally goes to waste, and this translates to wasting 1.47–1.96 global hectare (Gha) of arable land, 0.75–1.25 trillion m3 of water and 1 to 1.5 per cent of global energy (Aulakh et al 2013). Additionally, data from the International Monetary Fund (IMF) commodity prices shows a strong correlation between the price of crude oil and the FAO's Food Price Index.

Calls for nexus-framed policy approaches were backed by alarmist scenarios about the relationship between food, energy, water, and the climate. These approaches are accompanied by metaphors such as the "perfect storm", to quote John Beddington, the Chief Scientific Adviser to the British Government in 2009.

The nexus discourse needs to be situated within broader framings around multiple and interacting systems, food, water, energy but also other systems (see Beisheim 2013, who discusses widening the scope of the nexus to planetary boundaries). Scientific research has now claimed to identify potential critical thresholds and tipping points in the Earth system. A range of studies from system scientists argue that human activities drive multiple, interacting effects that cascade through the Earth system. Rockström et al (2009) state and quantify nine interacting "planetary boundaries" with possible threshold effects that manifest themselves at the planetary

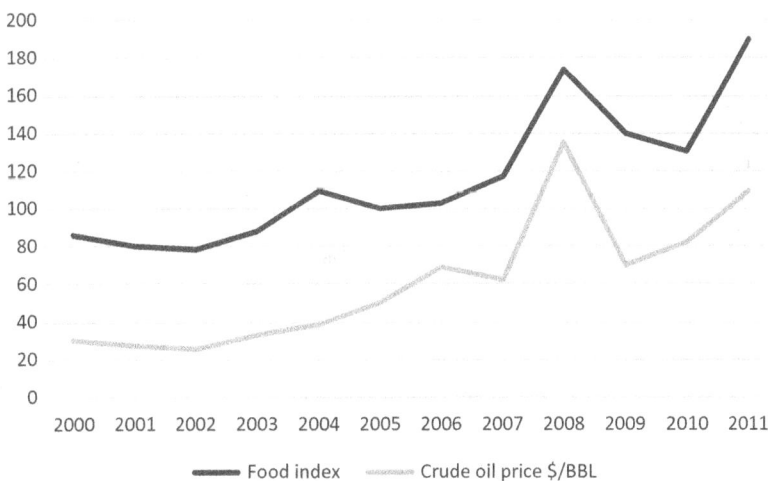

FIGURE 2.1 Food and crude oil prices, 2000–2011
Source: based on IMF – Primary Commodity Prices

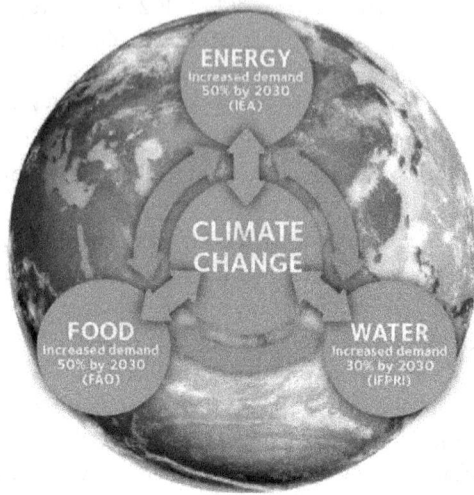

FIGURE 2.2 The perfect storm scenario
Source: Beddington 2009: Figure 7

level, possibly in a non-linear way, namely: climate impacts, ozone depletion, atmospheric aerosol loading, ocean acidification, global freshwater use, chemical pollution, land system change, biodiversity, and biogeochemistry (see also Steffen et al 2015). These "planetary boundaries" are however not fixed; they represent estimates of just how near the global human community can operate to an uncertainty zone around a potential threshold, without seriously challenging the continuation of the current state of the planet within which human settlements and cultures have flourished (Galaz et al 2012). However, drawing a "safe operating space for humanity" is a highly controversial and highly political project, as it echoes previous notions of "limits to growth" (Meadows et al 1972) and value judgments that are intrinsic to such estimates, which to date has included the privileging of expert scientific knowledge. While these efforts may bring a clearer focus on a range of biogeochemical processes and indicators at the planetary scale, the other side of the coin is they have also led to a new kind of "global conservationism" (see McAfee 2016) and a revitalisation of discourses with many apocalyptic neo-Malthusian crises narratives and concurrent debates focused on scarcity, population growth, and security. The current debates on the Anthropocene and planetary boundaries also miss how these processes play out in different parts of the planet or how and why these shifts affect diverse locales, species, or social groups in markedly different ways. This is why Jason Moore (2015), Andreas Malm (2015) and others reject the notion of the Anthropocene and instead focus on the socio-historical dynamics that led to the "Capitalocene", thus marking the unprecedented impact of capitalism, imperialism, and post-colonialism on humanity's relation with nature and its far-reaching consequences.

The backdrop of the oil crisis, and fuel shocks, created the presumption, at least amongst policy makers, that resources are "scarce" and they need to be managed and governed in a better way (Mehta 2010). The nexus' revival of limits to growth can therefore be seen as a policy response to a food and energy crisis and the fear of pessimistic scenarios about food, energy, and water supply and demand. One could therefore ask if the nexus is simply an economic project of scarcity management justified by claims of planetary boundaries and the survival of mankind. As put by Leese and Meisch (2015: 698): "all nexus conceptions share general perceptions of present and future crises and offer solutions for more efficient resource management within a green economy, thereby specifically calling for integrated solutions with regards to water, energy and food".

Scenario documents, whether in terms of food, water, or energy, are important not only for creating this sense of alarm and urgency in terms of perceived scarcity of resources, but also in terms of proposing how these can be managed. One can see that most scenario policy document tend to provide a narrative to manage economic structures through technological solutions, rather than question the distributive inequalities within the system. These framing have heavily influenced how nexus policy solutions have been designed (see for example WBCSD 2014).

Claims for resource scarcity are often evident in nexus framings, with crisis narratives created by some actors, epitomised by the WEF's Global Risks 2011 report:

> Shortages could cause social and political instability, geopolitical conflict and irreparable environmental damage. Any strategy that focuses on one part of the water-food-energy nexus without considering its interconnections risks serious unintended consequences.
>
> *WEF 2011b: 7*

The Bonn 2011 conference (under the headline The Water, Energy and Food-Security Nexus – Solutions for the Green Economy), which was intended as an input to the Rio +20 Earth Summit 2012 Conference, also underlined this competition over resources against the backdrop of urbanisation, growing population, and increasing levels of consumption, globalisation facilitating trade and investment, and resource degradation, and all amplified through climate change (BMU 2011; see also Bogardi et al 2012). The scarcity narrative of the nexus concept has also emerged in regional arenas. In Southeast Asia, for example, at the Asian Development Bank's (ADB) Greater Mekong Subregion 2020 conference on the nexus titled Balancing Economic Growth and Environmental Sustainability Focusing on Food–Water–Energy Nexus, scientific experts presented mainly technical assessments on how water resources are under growing pressure from agriculture, industry, energy production, and water extraction (Thapan 2012); and how food security, which is already insecure for millions across the region, is further at risk for a range of reasons including growing water scarcity (Rosegrant et al 2012).

Given the emphasis on scarcity, managing trade-offs has been seen as central to the nexus (Mushtaq et al 2009; ODI et al 2012; Rasul 2014). Largely adopting a systems approach, nexus thinking seeks to integrate sectors through making apparent the relationships between food, water, and energy systems, and addressing interconnected externalities (Howells et al 2013). Managing trade-offs is approached from the perspective of maximising benefits through minimising inefficiencies, reducing externalities, informing trade-offs through knowledge production, and identifying synergistic win–win scenarios where they exist. The Stockholm Environment Institute observes (Davis 2014: 2):

> In some cases, however, especially when resources are very scarce, a nexus analysis may not find a win–win option, but just difficult trade-offs (Weitz et al. 2014). The role of science in such situations is not to say what the "right" answer is, but to clarify the choices and ensure that all cross-sectoral impacts, externalities and trade-offs are known and understood. Participatory processes can also help ensure that vulnerable stakeholders have the information and access they need to advocate for themselves, and can foster dialogue across sectors and scales.

However, critical approaches to resource scarcity and the trade-offs go further and emphasise how scarcity can also be understood as a social phenomenon, shaped by market rules and other societal decisions. From this perspective, who experiences scarcity is the product of a politics of resource allocation that excludes particular groups from access (Mehta 2010; see also Hall, Hirsch, & Li 2011; Scoones et al 2014). Yet, across much of the nexus literature, the principles of distribution and governance of trade-off decisions – including who takes decisions and for whom – are inadequately problematised (Lele, Klousia-Marquis, & Goswami 2013). Leese and Meisch (2015) argue that most current framings of the nexus do not question the structured inequalities in the economy but rather confirm them. They view the nexus as a constructed security problem and a means for sustainability to be securitised; the nexus logic according to their view is plain and simple: we have to produce security through the economy and as such the survival of the economy must no longer be questioned. Foran (2015) critiques nexus systems approaches as under-theorised and under-politicised, in particular with regard to historical and relational considerations. It is clear that nexus thinking, if serious about attaining poverty reduction goals, needs to pay more attention to whose food, water, and energy that security is secured, including the means by which the needs of the marginalised will be prioritised (Allouche, Middleton, & Gyawali 2014; also Brauch 2011).

2.2. The nexus emerging from the World Economic Forum

The WEF first proposed the concept of the nexus in 2008. A year later, in 2009, the at-the-time UN Secretary-General, Ban Ki Moon, also addressed the WEF meeting at Davos and underlined the imperative of private sector participation to deal with these crises. This growing momentum led to a proliferation of special bodies within the WEF to deal with water issues, including the creation of a Global Compact CEO Water Mandate, the Global Agenda Council on Water

Security, the Global Water Initiative, and the 2030 Water Resources Group. It strongly links water security to economic growth in the following way:

> Water lies at the heart of a nexus of social, economic and political issues – agriculture, energy, cities, trade, finance, national security and human lives, rich and poor, water is not only an indispensable ingredient for human life, seen by many as a right, but also indisputably an economic and social good unlike any other. It is a commodity in its own right ... but it is also a crucial connector between humans, our environment and all aspects of our economic system.
>
> *WEF 2011a: 3*

The WEF's formulation of the nexus has primarily been driven by international private actors, who see the nexus – and subsequently also the concept of green economy – both as an opportunity and a constraint to their business. The WEF approach to the nexus stresses the business imperative and the need to prepare for investment scenarios in the near future. They underline that the economics of water is both compelling and challenging and that water security, economic development, and GDP are interlinked (WEF 2009). They thus argue that future global investments will be significantly driven by the consideration of water, and will become a mainstream theme for investors. This in turn holds policy implications, where the WEF seek to steer the flow of global water funds, as well as on clear-cut rules to price and manage water itself.

The business logic is as follows. To grow, economies should shift their water allocations away from farming and toward uses that deliver higher economic value per litre, especially energy production, industry, and manufacturing. Within this logic, governments therefore will have to make choices about the allocation of water between sectors (and, of course, are encouraged to pursue high value water uses). These shifts, at the same time, mean that they become more reliant on water use-efficient agriculture alongside food imports. To respond, the world system will need more trade flows in agriculture across more countries (WEF 2011a).

The WEF in particular emphasises market mechanisms as the solution to resource scarcity. Indeed, one of several explanations that the WEF gives for claims of a growing water scarcity and its risk to economic growth is the under-pricing of water as a resource; this, for example, has led to some regional water "bubbles" of agricultural prosperity, that in the long term are not sustainable, as water resources become depleted beyond the rate of replenishment. The report also argues that a weak international trade regime, and a complex arrangement of tariffs and subsidies, amplifies the cost of food shortage (WEF 2011a).

Another more recent report reiterated this approach from an energy perspective, emphasising that:

> Water must be seen as having its own economic value, and become a visible part of the equation to determine the most cost-effective energy options. No longer can power plants be evaluated just on fuel costs or capital costs. Smart

planning must take into account a richer model that includes life-cycle costs such as fuel collection, refinement and distribution as well as carbon and water costs.

WEF 2014: 9

In an earlier report, the WEF had referred to the "burst bubble" phenomenon as the concern that underpinned this nexus with the idea that, with growing resource scarcity and water insecurity, a business-as-usual approach was not feasible and in the long run is a threat to productivity and economic growth (WEF 2009a).

2.2.1. The WEF's resource realism

The WEF perspective has ushered in a new brand of resource realism (Wales & Winston 2012). It has provoked new public–private collaborations between international financial institutions (IFIs) and large transnational corporations such as Coca Cola, Nestlé, and SABMiller who want to harness the private sector's "comprehensive value-chain viewpoint" to tackle nexus governance and also advise governments, other corporates, and communities in nexus governance (Wales & Winston 2012). As the nexus gained salience in the WEF in 2011, they stressed four follow up measures: 1) A task force for data collection; 2) A major program of economic modelling of interdependencies; 3) New models for collaborations to help governments make policy changes; 4) Re-imagining institutions to improve governance (WEF 2011a). The operationalisation of the nexus is already underway, for example in debates on fracking or SABMiller's study on evaluation of malting barley in Rajasthan-India to determine value trade-offs across sectors (Reig, Luo, & Proctor 2014; Wales 2014).

The WEF approach to the nexus stresses the business imperative and the need to prepare for investment scenarios in the near future, with water and its relationships with food and energy systems amongst others as a key concern. As suggested by WEF, the investment scenarios and trade flows will be based on certain predictions of production and consumption. The complexity of decision situations, not least how unforeseen interactions between water, food, energy, and climate as well as overlapping – and often contradictory regimes – and the choices of other actors limit the predictability and applicability of such scenarios (Dequech 2006). Furthermore, the degree of uncertainty that veil the current climate change scenarios and models compound the already significant uncertainties. As suggested by Augier and Kreiner (2000), these choices do not necessarily represent optimal outcomes but "represent a gamble in an unanalysable world".

2.2.2. Scramble for resources

The scramble for resources is also evident in the 6th edition of the 2011 Global Risks report published by WEF (2011b), which made apparent that the "land-grab" phenomenon is a response to a larger structural and correlated global risk to water, energy, and food systems (see also Bizikova et al 2013). The use of market-based instruments for conserving what environmental economists refer to as "natural capital" have also led to the enclosure and securitisation of natural resources depriving communities of their traditional ownership rights and livelihood practices. Borras, Saturnino, and Franco (2012: 37) describe this as a process through which "national governments in 'finance-rich, resource-poor' countries are looking to 'finance-poor, resource-rich' countries to help secure their own food and especially energy needs into the future", resulting in many large-scale land deals related to grains and biofuels, very often driven by transnational corporations and very often with the active collaboration of the host national governments.

Thus the nexus thinking, in its simplistic form, might lead to the commodification of resources most readily or profitably monetised (perhaps for short-term gain), underplaying other long-term environmental externalities, such as biodiversity protection, pollution or climate change (Dupar & Oates 2012).

2.2.3. The Bonn 2011 conference

The German Federal government – which hosted the Bonn 2011 conference on the nexus – has been a key actor in pushing forward the nexus approach in global policy arenas. As put by the German Federal Ministry for Environment Nature Conservation and Nuclear Safety (BMU) and the Federal Ministry for Economic Cooperation and Development (BMZ) (2011), the nexus approach from their perspective identifies mutually beneficial responses and provides an informed and transparent framework for determining trade-offs and synergies that meet the demand without compromising sustainability. The Bonn paper underlined this competition over resources against the backdrop of urbanisation, growing population, and increasing levels of consumption, globalisation facilitating trade and investment, and resource degradation, and amplified through climate change (BMU 2011; see also Bogardi et al 2012). The nexus sectors – food, energy, and water – are thus understood to be interdependent and in need of integration; there is an emphasis on an intersectoral approach to breaking the silos between various sectors (Hoff 2011). "Nexus" thinking becomes a shorthand for this confluence of trends and need for explicit trade-offs in policy-making (Dupar & Oates 2012). Three guiding principles are proposed: investing to sustain ecosystem services; creating more with less; and accelerating access, integrating the poorest (Hoff 2011).

2.2.4. The global policy diffusion of the nexus

Both the WEF and the Bonn initiative have steered a new global policy debate over the nexus and its relationship with both achieving the Sustainable Development Goals (SDGs) and the Paris Agreement. As a result, the nexus has recently received increasing attention in international initiatives (e.g. SE4All) as well as support from the research/academic sector (e.g. IFPRI, DIE, SEI), governments (e.

g. German Development Cooperation), the private sector (e.g. WBCSD, AB InBev, Royal Dutch Shell, Coca Cola), and international development partners (e. g. REEEP OFID, IRENA, UNESCWA, FAO, EC, UNECE). The United Nations Economic Commission for Europe (UNECE) has established a water–food–energy–ecosystems nexus task force that looks at the nexus in selected trans-boundary river basins (De Strasser et al 2016). In addition, several international NGOs (in particular WWF and IUCN) have adopted the concept to promote the conservation of natural resources (see e.g. Ozment, Difrancesco, & Gartner 2015). The International Union for Conservation of Nature (IUCN) and the International Water Association (IWA) have initiated the Nexus Dialogue on Water Infra-structure Solutions to identify innovative approaches to the use of infrastructure, technology and finance to deal with challenges in the WEF nexus.

Most of the new policy reports tend to reinforce the vision promoted by the WEF. The World Business Council for Sustainable Development for instance published a report focusing on food security and the need for market approaches and pricing mechanisms (WBCSD 2014). In the same light, a report by Con-servation International (2018) focuses on the US drought in 2012 and the asso-ciated business risk and price volatility, and promotes market mechanism though public–private partnerships and productivity.

Other international policy actors take a more institutional approach. The FAO for example, which is a lead organisation in charge of advancing the High-Impact Opportunity (HIO) on the Water–Energy–Food Nexus (together with Germany) in the context of the Sustainable Energy for All initiative (SE4All), a UN action-focused global network, supported by partner organisations from governments, national and international organisations, businesses, and civil society organisations, emphasize on their side a cross sectoral and dynamic perspective to obtain the optimal management of trade-offs and the maximisation of overall benefits (FAO 2014). The FAO approaches the nexus by devising a systematic way to carry out a nexus assessment in a participatory way. It consists in a stepwise methodology, modelling water–food–energy interactions to support the decision-making system to assess the nexus context status, and also the performance of different technical and policy interventions against the country status. The nexus assessment can also be run independently in a non-participatory manner through a nexus rapid appraisal and, for this purpose, a number of key indicators and information sources are proposed (Flammini et al 2014). Another institutional approach is taken by WWF, SABMiller, and the Gesellschaft für Internationale Zusammenarbeit (GIZ) through the Water Futures Partnership programme. In a joint publication in 2014, WWF and SABMiller emphasize resilient government and business strategies to address the nexus, in particular arguing that the most resilient systems combine robust infrastructure, flexible institutions, and functioning natural capital. Their approach intends to share risks and coordinate decision making towards designing context specific policies (SABMiller/WWF 2014).

Overall, there is a clear political momentum around the nexus as exemplified in a 2012 poll of UN member states commissioned by the UN Secretary-General which showed that food, water, and energy were the top three priorities for the SDGs (Beisheim 2013: 2). By 2014, the nexus was the theme for World Water Day and being examined as the vehicle through which to deliver on the SDGs (Weitz et al 2014). "SDGS in 2017", a UN publication, noted the " … most commonly discussed set of interactions … " regarding the SDGs lie:

> … in the nexus between food, water and energy, as reflected in the links between SDG 2 [food], SDG 6 [water] and SDG 7 [energy], with potential conflict in water use for energy production and generating hydropower with residential and industrial water use and for irrigation for food production.
>
> *Nikolova et al 2017: 15*

However, despite this political momentum around specific policies, methodologies, and best case practices, the FAO (2018) in its report for the UN High-Level Political Forum on progress of SDGs considered that integrated approaches around the nexus had failed. It also highlighted that despite a strong gender dimension in the WEF sectors, gender aspects are often overlooked in the use of the WEF nexus approach. A particular issue for those promoting nexus approaches is that these approaches were not recognised into the outcome document of the 2012 United Nations Conference on Sustainable Development (despite major attempts through the Bonn conference described above). In fact, as emphasised by many experts, the UN 2030 Agenda for Sustainable Development, and in particular its 17 SDGs, still remain sectoral in their basic outlook. While a number of SDGs refer to other policy domains, this still remains somewhat random. The connections between many Goals are weak, and rarely structural or transparent. For instance, the Goal on hunger reduction and food security makes some connections to other issues such as equality, health, climate change, disasters, ecosystem protection, and infrastructure. Yet this does not explicitly refer to the interconnections with water and energy, among many other potential connections. Similarly, the Goal on ensuring the availability and sustainable management of water and sanitation for all makes no explicit connection to food or climate change (Boas, Biermann, & Kanie 2016).

2.2.5. The Green Washington consensus

The dominant framing of the nexus is very much aligned with green economy debates (Pearce, Markandya, & Barbier 2013). The business imperative which is driven by increasing resource and productive efficiency is fairly evident. In the wake of the financial crises, the green economy was seen as an entry point to promote green investments (Ocampo 2011). Though green economy was originally conceptualised as a set of economic tools to operationalise the SDGs (Jacobs 2012), the concept has been broadened and has become deeply contested over time (Benson & Greenfield 2012). Hoff (2011) in a highly-cited paper on the nexus prepared for Bonn 2011, describes the green economy and its relationship with the nexus as:

> ... an economy that results in improved human wellbeing and social equity, while significantly reducing environmental risks and ecological scarcities [...]. In Green Economy natural capital is valued as a critical economic asset as a provider for benefits for the poor [...] it is the nexus approach par excellence.
>
> *Hoff 2011: 6*

On similar lines, the starting point for the OECD's deliberations on "Green Growth" is the risk of climate change and depletion of natural resources such as unchecked biodiversity loss, overfishing, and the growing scarcity of land and water. The OECD report (2011: 4) states that "We need green growth because risks to development are rising as growth continues to erode natural capital". The way forward should be characterised by 1) increased productivity; 2) increased (technological innovation); and 3) stimulating new markets.

The United Nations Economic and Social Commission for Asia and the Pacific (UNESCAP), in a position paper published in 2013 on the WEF nexus in Asia and the Pacific region, acknowledging the "dearth of studies on the interconnections between water–food–energy in the Asia Pacific region" recommends the wider region to:

> [e]mbrace green economy as a new policy goal and pursue 'low carbon, resource efficient, and socially inclusive' development strategies as espoused in the United Nations Conference on Sustainable Development (UNCSD or Rio+20) .... The world needs to find profitable market-oriented solutions to nexus challenges ....
>
> *UNESCAP 2013: 49*

The solutions proposed by the Green Economy are often about increasing productivity through resource efficiency or availability of resources. Like the nexus, the green economy focuses on supply side challenges rather than questioning the factors which increase the demand for resources, and questioning the patterns and inequities in consumption and lifestyle (Unmüßig, Sachs, & Fatheuer 2012). It has promoted, for example, "clean energy", although these options such as "clean coal", large hydropower dams, large solar farms etc. raise many complex socio-environmental and social justice issues. For example, the rising production of biofuel crops have been linked to deforestation, landgrabs, and competing uses of land with agriculture, including small-holder agriculture (Borras, McMichael, & Scoones 2010).

The green economy pushes for a commodification of natural resources, which provides for some a way to deal with resource scarcity linked to the nexus by putting a price on natural resources whereby nature is made subject to financial speculation. It promotes monetisation and financialisation of nature, which is a set of three processes: 1) creation of commodities or marketable goods through market or valuation techniques; 2) introduction of commercial principles such as efficiency, accounting or cost-benefit assessment and profit maximisation; and 3) creation of tradable instruments such as environmental credits and offsets (Huff 2015).

The strategy detailed is essentially an ecological modernisation project, orientated around a market-based approach of resource productivity, efficiency, and technology. Not all endorse the concept of Green Economy; the World Social Forum has called it "the Green Washington Consensus", stating "this latest phase of capitalist expansion seeks to exploit and profit by putting a price value on the essential life-giving capacities of nature" (Working Group on Green Economy 2012). At Rio+20, the green economy was also contested by developing countries worried it may prove a vehicle by which, "industrialized countries slip out of their commitments to promote and fund sustainable development, while imposing new forms of environmental conditionality on resource use" (Conca 2015a: 169). Currently, the framing of the nexus is very top-down, often North to South linked to external interests, and outsider-generated managerial solutions.

2.2.6. Policy diffusion of the nexus: Case study of Southeast Asia

While much more prominent in global-level policy arenas, the nexus is also diffusing to shape regional and national-scaled policies. For example, in mainland Southeast Asia a range of international organisations and civil society, academics, and high-income country donors, working with the region's governments and politicians, are translating and diffusing the nexus concept through their research, programming, and policy recommendations.

A variety of initiatives have emerged, including those led by the Asian Development Bank (ADB) geared towards shaping its investment agenda (ADB 2012b), by regional research and policy platforms such as the Challenge Program on Water and Food – Mekong (CPWF-Mekong 2013) and its more recent iteration as the Water–Land–Ecosystem Mekong (WLE-Mekong) Program, and by intergovernmental organisations such as the Mekong River Commission (Bach et al 2012). On the ground, some researchers have also taken the nexus as a heuristic framework by which to force thinking on the relationship between food, energy, and water trade-offs and operationalise the nexus in participatory planning and decision-making processes (Krittasudthacheewa et al 2012; Smajgl & Ward 2013). Despite these activities, the region's social movements and local and national civil society have yet to seriously discuss or adopt the nexus concept as framed in global and regional policy circles.

Since 2011, there has been a growing volume of meetings and reports addressing the nexus, and thus the concept itself has grown in prominence. Dore, Lebel, and Molle (2012: 26) observe that it is within the nexus discourse that "many actors see a logical, sectoral entry point for themselves in compelling new, multi-sector, interdisciplinary and transboundary deliberations". Three broad types of organisations have engaged in the nexus debate:

- *Investment/Lending organisations*: ADB; World Bank.
- *Sustainable development organisations and research institutes*: Mekong River Commission (MRC), the CGIAR CPWF-Mekong, the Stockholm International Water Institute (SIWI), United Nations Economic and Social Commission for

Asia and the Pacific (UNESCAP), United Nations Environment Programme (UNEP), IWA, International Water Management Institute (IWMI), the Stimson Centre, the Commonwealth Scientific and Industrial Research Organisation (CSIRO), and the Stockholm Environment Institute (SEI).

- *Conservation organisations*: IUCN; and the Worldwide Fund for Nature (WWF).

Major donors funding the above organisations include the governments of Australia, Finland, Denmark, Germany, Sweden, the United Kingdom, and the United States.

Organisations promoting the nexus and their donors have commissioned research, supported networks of government policy makers, academics and civil society, and organised conferences in promoting and deliberating the nexus. The first major conference on the nexus in Southeast Asia was the 1st Mekong Forum on Water, Food and Energy, held in Phnom Penh, Cambodia in December 2011 and organised by the CGIAR CPWF-Mekong, with funding from Australia Aid. Since 2011 large conferences have been held almost annually in various cities of Southeast Asia, under CPWF-Mekong, and subsequently WLE-Mekong. These conferences, which have grown in size and scope, are convened as multi-stake-holder dialogues and regularly draw upon the nexus as a framing concept to draw senior officials from an array of government agencies responsible for water, food, energy, and the wider economy.

The CPWF-Mekong/WLE-Mekong program, together with research network called the Mekong Program on Water, Environment and Resilience (M-POWER), have built a research-driven epistemic network[3] around water govern-ance in the region that has increasingly researched and deliberated the nexus at their conferences and other convened dialogues (Dore 2014; Dore, Lebel, & Molle 2012). These networks are formed of international organisations, regional and international researchers and civil society, regional governments, and mainly high-income country donors. Thus, the nexus concept has also spread into other con-ferences and policy forums that draw upon active members of these networks. Many research projects to date, however, may be better characterised as multi- or inter-disciplinary and cross-sectoral in investigating particular nexused relationships, rather than specifically engaging the nexus as taken "off-the-shelf" of global policy.

Governments have also broadly engaged with the nexus, the most prominent being via the MRC, an intergovernmental transboundary river basin organisation for the Mekong Basin that until recently drew its budget principally from high-income country donors. In May 2012, the MRC hosted the Mekong2Rio Inter-national Conference on Transboundary River Basin Management that convened experts from fourteen major transboundary river basins and sixteen related inter-national organisations to reflect on the nexus-approach (Bach et al 2012). By hosting the conference, the MRC also placed the Mekong basin as a key focal object in the discussion of nexused transboundary river management globally. The Mekong2Rio conference's conclusions fed into the global Rio+20 process in June

2012 (Bach et al 2012: 60–61). Subsequently, in April 2014, another major nexus-framed international conference was hosted by the MRC (Bach et al 2014),[4] from which a nexus-framed message was delivered in person to the region's highest-level political leaders who attended the accompanying Second MRC Summit:

> In order to collectively benefit from the opportunities [of the nexus perspective], transboundary agreements and institutions develop and need to adapt to changing environments. For these to work effectively, a combination of political will, technical cooperation and an inclusive process is required.

This demonstrates a significant shift in framing from the message delivered to the region's political leaders at the First MRC Summit held four years earlier, which was framed around the implementation of transboundary IWRM (MRC 2010).

On a different tack, the ADB also promoted the nexus through a major conference in Bangkok in February 2012 under its Greater Mekong Subregion (GMS) programme, which is principally geared towards accelerating regional economic integration through infrastructural investment, institutional reform, and capacity building (ADB 2012a). At the conference, experts presented on the nexus principally to senior government officials, representatives from the private sector, donors, and select civil society. The conference provided input to the GMS programme, with the conference proceeding preface stating:

> The progress [of the GMS programme] is reflected in terms of improvements in infrastructure connectivity, promotion of trade and investment, stimulation of economic growth, and reduction of poverty. However, such progress has not been without some adverse impacts on the environment .... Many presentations ... focused on deepening the awareness and understanding of the nexus as a basis for transition to climate-resilient and green economic pathways of development .... ADB is committed to play its part in assisting countries in the subregion to achieve this goal by mobilising additional financial resources and developing new knowledge products.[5]

The ADB's approach to the nexus, working principally with the region's governments, thus reflects an approach to define its investment strategy with the conference outcomes framed in terms of promoting economic growth and a green economy, rather than giving explicit consideration to social justice. The ADB has launched several additional major reports on water and the nexus, most notably "Thinking about Water Differently: Managing the Water–Food–Energy Nexus" (ADB 2013). This report, whilst wide-ranging, emphasises how economic water scarcity in Cambodia, Laos, Vietnam, and Myanmar could be addressed through "improving supply side infrastructure", whilst also addressing demand-side factors (including through water pricing), strengthening governance, and building new institutions (ADB 2013: vi–vii), thus again explicitly linking the nexus to the ADB's investment strategy.

Overall, only a limited number of international NGOs and policy think tanks have been drawn to the nexus in the region to date. The Stimson Institute, in their report "Mekong Turning Point: Shared River for a Shared Future", frame their subsequent analysis on the risks posed by plans for mainstream dams on the Mekong River with the opening sentence: "In no part of the world does the increasingly critical nexus of water, food, and energy have more immediate relevance than the Mekong River, a transboundary resource shared by China and five Southeast Asian countries" (Cronin & Hamlin 2012: 1). Another example is the WWF's Mekong Nexus Project initiated late in 2014 designed to "research key links, conflicts and positive synergies between conservation of biodiversity, responses to climate change, and supply of energy, food and water" (WWF 2014). On the other hand, national and local civil society groups have rarely explicitly utilised the nexus as a framing for their work to date.

Finally, a number of global-level nexus initiatives have also sought to gather experience from Southeast Asia to both promote the nexus in Southeast Asia and project the region back into global policy arenas. In addition to the CPWF program and Mekong2Rio conferences mentioned above, most notable has been the IWA–IUCN Nexus Dialogue on Water Infrastructure Solutions that held three "regional dialogues", including in Bangkok in March 2014, and subsequently a global synthesis meeting in Beijing in November 2014 (GWP-China, IWA, & IUCN 2014). The IWA in its framing of the dialogue emphasised how the nexus "has led to new demands for water infrastructure and technology solutions"[6] whilst IUCN has sought a framing emphasising "natural infrastructure" (see Krchnak, Mark, & Deutz 2011).

Several initiatives in Southeast Asia have sought to operationalise the nexus through their field-based work. An innovative research project titled Exploring Mekong Region Futures, led by the Commonwealth Scientific and Industrial Research Organization (CSIRO) (2009–2013) used the nexus as a heuristic tool in a regional "Delphi" assessment, together with five local case studies to explore a range of development scenarios and alternative futures (Smajgl & Ward 2013). For example, the Northeast Thailand Futures study led by the SEI worked with farmers, local government, academics, and others to explore a range of scenarios related to rice, sugarcane, cassava, and rubber production in the context of rising demands for energy (including biofuels) and food in the Huai Sai Bat (HSB) sub-basin of the Chi River (Krittasudthacheewa et al 2012). The learning-orientated research design and involvement of Thailand's National Economic and Social Development Board facilitated the inclusion of a form of nexus-concept into the 11th five-year National Economic and Social Development Plan for Northeast Thailand.

This section has mapped how the nexus has spread throughout mainland Southeast Asia from global-level policy conceptualisation to within regional policy circles that have included international and regional organisations, academic networks, and civil society, national politicians and government officials, and high-income country donors. There is some indication of the spreading and embedding of the nexus in the region ranging from the presentation of the concept to the region's top political leadership at various policy forums, to the involvement of government, regional academics, and others in dialogue meetings and conferences, to the growing volume of academic and policy-orientated research. The nexus is yet to be extensively grounded, however, into national policies and practices, and broad-based local demand for nexus-framed policies is currently limited. Molle (2008: 143) suggests that a "snowballing effect" results in a growing number of actors promoting and implementing a particular nirvana policy concept, such that it is "gradually established as a consensual and controlling idea". In the case of the nexus, it still remains to be seen whether this will occur in Southeast Asia.

2.3. Critique and dynamic understanding of the nexus

The current framing of the nexus has problems as it tends to draw simplistic and often apolitical causal relations between availability and access. For example, fears expressed in the WEF-framed nexus' around potential food crises portray a simple causality between food and water availability and reduction in hunger or improved access to water (Allouche et al 2014). Yet, the relationship between availability and access is often mediated by monetary (resources available to access) and non-monetary factors (inequitable power relations in the given context, institutional arrangements) that are critical in providing access.

The logic of optimisation, embedded within the nexus thinking, also has limits as it treats resource allocation in a perfect equilibrium model (Allouche, Middleton, & Gyawali 2015). This can encourage the commodification of resources, neglecting other issues such as environmental externalities, poverty alleviation, and everyday realities at the local level (see Dupar & Oates 2012). For many rural communities, food, water, and energy has never been conceptually separated in the way that experts have sought to understand them. Indeed, it may be that the water–food–energy nexus is the (re)discovery by experts working in silos of what practicing farmers and fishers already know. People living in rural communities engage in what Batterbury (2001) has called a "productive bricolage", juggling use of different community resources depending upon tenure and access, season, gender, labour availability, and markets. At different times, they may be a farmer, a fisherfolk, a forager, and/or a forester, as well as increasingly a migrant labourer. When one critical resource or ecosystem is affected by the externalities of large resource development, rural people tend to diversify and devote more of their energies to using other resources, thus re-creating new social landscapes through changing practices of property, resource access, and use. Rural peoples' livelihoods span the resource sectors that modern planning tends to divide and compartmentalise (as involving aquatic and forestland management). Valuing nature as a particular kind of "natural capital" and the regimes of financialisation and commodification that accrue with them, may not only produce inequalities related to the reallocation of access and control, but would also mean disrupting these "local nexuses" which are often more attuned to addressing complexity and uncertainty in a particular situational context.

In addition, in framing the nexus around crisis, a space for appropriation is opened up, often linked to a partial enclosure of previously shared, regional commons (a form of "green grabbing"). Investment imperatives linked to notions of food, energy, or water "security" drive such appropriations by the private sector, supported by national political interests. From the perspective of social justice, as discussed above, scarcity may result from inequitable allocation and access, rather than due to an absolute scarcity (Mehta 2010). Others have also highlighted that water scarcity limiting economic growth is place specific and may not be the general case (Barbier 2004). For example, Brown and Lall (2006) suggest that it is rainfall variability in particular that is a significant factor shaping economic growth, rather than a generalised water scarcity.

We live in a complex world. There are a multitude of interactions between the material, social, and the biophysical world. Nexus thinking may be described as an attempt to render these interactions visible and amenable to policy making (cf. Dupar & Oates 2012), but there are clear limits to such a framing and their corresponding solutions. As discussed above, it is difficult to predict and capture the dynamics of this interaction as they are ridden with uncertainties – the unknown unknowns – more so in the context of climate change. There is a tendency towards the idea of stability, to bring the "situation under control" with expressions such as "halting" or "curbing" climate change. There is an illusion that this "stability" might be achieved through manipulating a few variables, out of the millions of interlinked and dynamic factors, which govern the world's climate. As expressed by Philip Stott already in relation to ecological systems:

> Our ecological language is suffused with a desire for "stability" and "safety" (the so-called "precautionary principles"), whereas in reality all is Heracleitan flux, and we can never "step into the same river twice." We are trying to replace human flexibility and adaptation by god- like control and statis, and it will not work.
>
> *Stott 1998: 1*

Leach, Scoones, and Stirling (2010) show how recent understanding of ecological systems has shifted from seeing nature as "in balance", a rather mechanistic approach, to recognising ecological systems as in a dynamic non-equilibrium with potentially non-linear responses and multiple stable states. This dynamism becomes more complex as ecological, social, economic, technological, and political systems interact. The hydrological cycle, for example, is a highly dynamic system – ecologically, socially, and technologically (Mehta et al 2007) – and will become *even more* dynamic under the conditions of climate change.

Narratives prioritise different aspects of systems dynamics and propose different strategies to deal with them.

- **Stability:** If a system is assumed to move along an unchanging path, the strategy may be designed to exercise control
- **Resilience:** If limits to control are acknowledged, the strategy might be to resist shocks in a more responsive way

- **Durability:** If a system may be subject to stresses and shifts over time, interventions may attempt to control the potential changes
- **Robustness**: Strategies that embrace both the limits to control and an openness to enduring shifts

Leach, Scoones, and Stirling (2010) argue that often "static" thinking (big, technically driven, managerial solutions that seek to control variability) rather than "dynamic" thinking (adaptive, reflexive thinking and "clumsy" solutions that accommodate and respond to change) often prevails, resulting in failed outcomes for projects and policies. In much literature and policy discussion to date, the framing of the nexus has been very top down, often linked to external interests, and outsider-generated managerial control-type solutions. Solutions fronted by the nexus often prioritise particular causal loops which are over simplistic and neglect the political economy of these interactions especially around the questions of resource access and use. Nexus thinking tends to propose solutions where closed stability-oriented technological solutions are fronted. There are thus inherent dangers that this "closed" framing may render local actions and needs invisible or peripheral as they get pushed out of the policy discourse. It often fails to ask the critical question of: who benefits from these improvements and at whose or at what costs?

Thus, conceptually, this book aims to introduce non-equilibrium thinking to the nexus. It requires a transition from analytical assumptions of equilibrium thinking, centred on linearity, predictability, homogeneity, and simplification to ones that encompass non-linearity, complexity, uncertainty, ambiguity, and surprise (Leach, Scoones, & Stirling 2010). In light of myriad physical uncertainties and social vulnerabilities that different societies in varied ecosystems face, it strongly suggests that a nexus approach actually requires pursuing plural pathways that work in bottom-up ways and that are more attuned to local systems of resource availability as well as uses

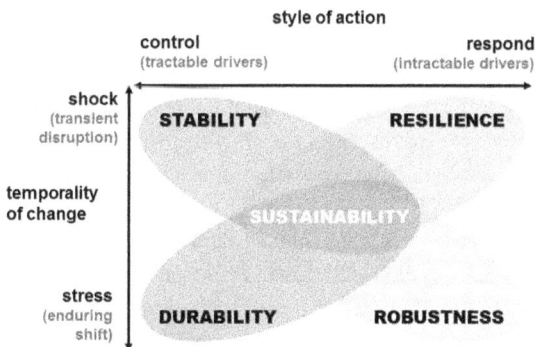

FIGURE 2.3 Dynamic properties of sustainability
Source: Leach, Scoones, & Stirling 2010: 62

and consumption. In this light, reversing the gaze and thinking of systems as dynamic and open-ended becomes crucial. Priorities can also look very different at the local level. We need to think of better ways to understanding and work with alternative development choices which privilege local and diverse ways of knowing, as we work across scales to understand these ecological, social, and technical interactions.

Both cultural and ecological theories may help to sharpen our understanding of dynamic security by integrating and accepting inevitability of change in planning. The Theory of Plural Rationalities (also known as Cultural Theory) argues with its "theory of surprise" that different ways of organising are based on different ways of understanding how the world is and how it supposed to work.[7] When the world (both physical as well as the social that exploits the physical in different ways) happens to behave in a way not anticipated by the particular way of organising (but advocated to do so by other ways of organising opposed to that particular way) there is surprise and re-adjustment both of worldview and behaviour. Thus *surprise* – the outcome of the ever-widening discrepancy between the expected and the actual – is of central importance in dislodging people from their ways of organising and helps us onto a "theory of change" that makes change permanent and intrinsically forever in dynamic flux (Thompson 2008). As argued by Gujit (2008: 287):

> the predominantly positivist and "development-as-project" vision that guides such monitoring is inconsistent with the emergent and non-linear nature of institutional change that occurs through "messy" partnerships and that is increasingly central in rural development and resource management. It is also inconsistent with the everyday reality of monitoring as a continual informal dialogue among development actors, not bound by official monitoring procedures and protocols.

Pursuing plural pathways to accommodate complexity and (climate) uncertainty would also require a shift in governance from "stability and durability thinking" that tends towards control approaches and the construction of (large) man-made structures and towards incorporating "resilience and robustness thinking" where the limits to control are acknowledged and adaptive solutions that incorporate plural solutions are pursued. In light of myriad physical uncertainties and social vulnerabilities that different societies in varied ecosystems have to cope with, what "robustness" would mean is not one perfectly optimised, large-scale solution but what has been called "many ten percent solutions" (NCVST 2009) amenable to action by different social solidarities.

Drawing on Cultural Theory, these plural solutions might be considered as "clumsy solutions" that benefit from the deliberative interaction of multiple worldviews on perceptions of problems and associated risk, ways of organising social relations, and ultimately generating creative, new solutions and alternatives and plural policy responses (Verweij & Thompson 2006; See also Gyawali 2009). Indeed, complex water–food–energy nexus problems are so-called wicked problems to which there is no easy definition and no easy solution (Lach, Rayner, & Ingram 2005; Rittel & Webber 1973). Dynamic sustainability approaches and plural, clumsy solutions – given their reflexive and interdisciplinary perspectives – are particularly apt to be applied to such challenges. Furthermore, plural solutions to water and food and energy challenges offer the potential of better outcomes in terms of both effectiveness and social justice outcomes.

In this book, following a pathways approach,[8] we seek to broaden out the inputs to planning processes and appraisal methods, and open up the outputs to decision-making and policy to recognise the different pathways to sustainability around the nexus.[9] The book highlights diversity, distribution, and direction, namely: the need to move away from social and technological inflexibility indicators[10]; the need to recognise those excluded and marginalised from preferred solutions; and the need to recognise alternative pathways. The way forward for nexus thinking thus begins with acknowledging the value of plural frames and solutions, and recognising the deeply political nature of resource use and access.

Notes

1 These include CH2M HILL, Cisco Systems, the Coca-Cola company, the Dow Chemical company, Halcrow Group, Hindustan Construction Company, McKinsey, Nestle, PepsiCo, Rio Tinto Group, SABMiller, Standard Chartered Bank, Syngeta AG, Unilever. See www3.weforum.org/docs/WEF_WI_WaterSecurity_WaterFoodE nergyClimateNexus_2011.pdf.

2 www.newsecuritybeat.org/2008/03/rising-food-prices-destabilizing-dozens-of-countries/.

3 Haas and Haas (1990: xviii) define an epistemic community as: "knowledge-based groups of experts and specialists who share common beliefs about cause-and-effect relationships in the world and some political values concerning the ends to which policies should be addressed".

4 The International Conference on Cooperation for Water, Energy and Food Security Under Climate Change in the Mekong Basin, 2–3 April, 2014, Ho Chi Minh City,

5 www.gms-eoc.org/uploads/wysiwyg/events/GMS2020-Final-Proceedings/Cover-Foreword-Opening%20Remarks-Inauguration.pdf.

6 www.iwawaterwiki.org/xwiki/bin/view/Articles/NexusDialogueonWaterInfrastructure Solutions-BuildingPartnershipsforInnovationinWaterEnergyandFoodSecurity.

7 Cultural Theory's four ways of organising (four social solidarities) are: bureaucratic hierarchism, market individualism, activist egalitarianism, and the fatalism of non-galvanised masses, the voters, or consumers. The worldview of each way of organising is only partly right and never wholly wrong but has to constantly be in battle with other ways of organising as the world keeps dishing out unexpected surprises. See Thompson, Grenstad, & Selle 1999.

8 Development pathways, defined as "the particular directions in which interacting social, technological and environmental systems co-evolve over time" (Leach, Scoones, & Stirling 2010: xiv), can be either sustainable or unsustainable. Pathways are historically contingent and often become "locked-in" for many reasons including due to past decisions and sunk investment, alongside technological, social, and political inflexibility. The pathways approach has been applied to a range of issues, including: climate change; energy; pandemic disease; water scarcity; hunger; poverty; and inequality

9 "Broadening out inputs" to planning processes and appraisal methods includes: participatory engagement; extended scope to include multiple criteria and scales; an acceptance of a diversity of knowledges; the need to acknowledge uncertainty; and the importance of addressing issues of rights, equity, and power. "Opening up the outputs" to decision-making and policy includes: giving serious consideration to a range of options and possible alternatives; and a move towards more adaptive, deliberative and reflexive forms of governance and political engagement (Leach, Scoones, & Stirling 2010: 100).

10 One set of "indicators of inflexibility" we play with is provided by Thompson (1994) building on Collingridge. His four indicators of technical inflexibility are: large scale, long lead time, capital intensive, and meeting major infrastructure needs early on; and the four of social inflexibility are: "single mission" outfits, closure to criticism, hype (as in "If we do not cover the Himalayas with trees, Bangladesh will forever sink beneath the waves"), and hubris (often in the form of over-confidence as to what the future holds, or categorical certainty that "There is no alternative").

References

ADB (2012a). *Asian Development Outlook 2012: Confronting rising inequality in Asia*. Manila: ADB.

ADB (2012b). *International Conference on GMS 2020: Balancing economic growth and environmental sustainability, focusing on food-water-energy nexus*. Bangkok, Thailand: ADB.

ADB (2013). *Thinking about water differently: Managing the water-food-energy nexus*. Mandaluyong City, Philippines: ADB.

Allouche, J. (2011). The sustainability and resilience of global water and food systems: Political analysis of the interplay between security, resource scarcity, political systems and global trade. *Food Policy*, 36: S3–S8.

Allouche, J. (2016). The birth and spread of IWRM: A case study of global policy diffusion and translation. *Water Alternatives*, 9(3), 412–433.

Allouche, J., Middleton, C., & Gyawali, D. (2014). Nexus nirvana or nexus nullity? A dynamic approach to security and sustainability in the water-energy-food nexus. STEPS Working Paper 63, Brighton: STEPS Center. Downloadable at: http://steps-centre.org/wp-content/uploads/Water-and-the-Nexus.pdf.

Allouche, J., Middleton C., & Gyawali, D. (2015). Technical veil, hidden politics: Interrogating the power linkages behind the nexus. *Water Alternatives*, 8(1), 610–626. Downloadable at: www.water-alternatives.org/index.php/alldoc/articles/vol8/v8issue1/277-a8-1-1/file.

Augier, M. & Kreiner, K. (2000). An interview with James G. March. *Journal of Management Inquiry*, 9(3), 284–297.

Aulakh, J., Regmi, A., Fulton, J., & Alexander, C. (2013). Food losses: Developing a consistent global estimation framework. In *Agricultural and applied economics association annual meeting*, August (pp. 4–6).

Bach, H., Bird, J., Clausen, T. J., Jensen, K. M., Lange, R. B., Taylor, R., Viriyasakultorn, V., & Wolf, A. (2012). *Transboundary river basin management: Addressing water, energy and food security*. Lao, PDR: Mekong River Commission.

Bach, H., Glennie, P., Taylor, R., Clausen, T. J., Holzwarth, F., Jensen, K. M., Meija, A., & Schmeier, S. (2014). *Cooperation for water, energy and food security in transboundary basins under changing climate*. Lao PDR: Mekong River Commission.

Barbier, E. B. (2004). Water and economic growth. *Economic Record*, 80, 1–16.

Batterbury, S. (2001). Landscapes of diversity: A local political ecology of livelihood diversification in south-western Niger. *Ecumene*, 8(4), 437–464.

Beddington, J. (2009). *Food, energy, water and the climate: A perfect storm of global events?* London: Government Office for Science, http://webarchive.nationalarchives.gov.uk/20121212135622/ (accessed on 5 April 2014).

Beisheim, M. (2013). The water, energy & food security nexus: How to govern complex risks to sustainable supply. *SWP Comments*, 32, 1–8.

Benson, E. & Greenfield, O. (2012) Surveying the "green economy" and "green growth" landscape. Draft for consultation. Green Economy Coalition.

Benson, T., Minot, N., Pender, J., Robles, M., & von Braun, J. (2008). Global food crises: Monitoring and assessing impact to inform policy responses. *International Food Policy Research Institute, Food Policy Report*, September 2008, Washington DC: IFPRI.

Bizikova, L., Roy, D., Swanson, D., Venema, H. D., & McCandless, M. (2013). *The water-energy-food security nexus: Towards a practical planning and decision-support framework for landscape investment and risk management*. Manitoba, Canada: The International Institute for Sustainable Development.

BMU (2011). Thematic profile. Paper presented at the Bonn2011 Conference: The water, energy and food security nexus solutions for the green economy, Bonn, 18 November 2011.

Boas, I., Biermann, F., & Kanie, N. (2016). Cross-sectoral strategies in global sustainability governance: towards a nexus approach. *International Environmental Agreements: Politics, Law and Economics*, 16(3), 449–464.

Bogardi, J. J., Dudgeon, D., Lawford, R., Flinkerbusch, E., Meyn, A., Pahl-Wostl, C., Vielhauer, K., & Vörösmarty, C. (2012). Water security for a planet under pressure: interconnected challenges of a changing world call for sustainable solutions. *Current Opinion in Environmental Sustainability*, 4, 35–43.

Borras, S. M. Jr. & Franco, J. (2012a). A "land sovereignty" alternative? Towards a peoples' counter-enclosure. TNI Agrarian Justice Programme Discussion Paper. Amsterdam: Trans National Institute (TNI).

Borras, S., McMichael, P., & Scoones, I. (2010). The politics of biofuels, land and agrarian change, Editors' Introduction. *Journal of Peasant Studies*, 37(4), 575–592.

Borras, S. M. Jr., Saturnino, M., & Franco, J. (2012b). Global land grabbing and trajectories of agrarian change: A preliminary analysis. *Journal of Agrarian Change*, 12(1), 34–59.

Brauch, H. G. (2011). Concepts of security threats, challenges, vulnerabilites and risks. In H. G. Brauch, Ú. Oswald Spring, C. Mesjasz et al. (eds). *Coping with global environmental change, disasters and security*. Hexagon Series on Human and Environmental Security and Peace. Berlin, Heidelberg, and New York: Springer-Verlag.

Brown, C. & Lall, U. (2006). Water and economic development: The role of variability and a framework for resilience. *Natural Resources Forum*, 30, 306–317.

Conca, K. (2015). *An unfinished foundation: The United Nations and global environmental governance*. Oxford: Oxford University Press.

Conservation International (2018). *The Energy / Food / Water Nexus, Resources, Vol. 2*. www.conservation.org/publications/Documents/BSC_Resources_vol2.pdf (accessed 3 October 2018).

CPWF-Mekong (2013). Mekong Forum on Water, Food and Energy. November 19–21, 2013, Ha Noi, Vietnam. Vientiane: CGIAR Challenge Program on Water and Food, M-POWER, IWRP, CGIAR Research Program on Water, Land and Ecosystems, Australian Aid, and IFAD.

Cronin, R. & Hamlin, T. (2012). *Mekong turning point: Shared river for a shared future*. Washington, DC: Stimson Center.

Davis, M. (2014). *Managing environmental systems: The water-energy-food nexus. Research Synthesis Brief*. Stockholm: Stockholm Environment Institute (SEI).

Dequech, D. (2006). The new institutional economics and the theory of behaviour under uncertainty. *Journal of Economic Behavior & Organization*, 59(1), 109–131.

De Strasser, L., Lipponen, A., Howells, M., Stec, S., & Bréthaut, C. (2016). A methodology to assess the water energy food ecosystems nexus in transboundary river basins. *Water*, 8, 59.

Dore, J., Lebel, L., & Molle, F. (2012). A framework for analysing transboundary water governance complexes, illustrated in the Mekong Region. *Journal of Hydrology*, 466–467, 23–36.

Dore, J. (2014). An agenda for deliberative water governance arenas in the Mekong. *Water Policy*, 16(S2), 194–214.

Dupar, M. & Oates, N. (2012). *Getting to grips with the water-energy-food "nexus"*. London: Climate and Development Knowledge Network. Available online: http://cdkn.org/2012/04/getting-to-grips-with-thewater-energy-food-nexus/ (accessed on 10 July 2015).

FAO (2008). *Water for the rural poor: Interventions for improving livelihoods in sub-Saharan Africa.* Rome: Food and Agriculture Organization (FAO).

FAO (2011). *The state of the world's land and water resources for food and agriculture.* Summary Report. Abingdon and New York: Earthscan and FAO. www.fao.org/docrep/015/i1688e/i1688e00.pdf (accessed 10 March 2017).

FAO (2014). *The Water-Food-Energy Nexus: a new approach in support of food security and sustainable agriculture.* Rome: FAO.

FAO (2018). Water-Energy-Food Nexus for the Review of SDG7, Policy Brief 9, UN High- Level Political Forum (July 2018).

Flammini, A., Puri, M., Pluschke, L., & Dubois, O. (2014). Walking the nexus talk: assessing the water-energy-food nexus in the context of the sustainable energy for all initiative. Environment and Natural resources Management Working Paper, 58. Rome: FAO.

Floyd, R. (2010). *Security and the environment: Securitisation theory and US environmental security policy.* Cambridge, UK: Cambridge University Press.

Foran, T. (2015). Node and regime: Interdisciplinary analysis of water-energy-food, nexus in the Mekong region. *Water Alternatives*, 8(1), 655–674.

Galaz, V., Biermann, F., Crona, B., Loorbach, D., Folke, C., Olsson, P., & Reischl, G. (2012). Planetary boundaries: exploring the challenges for global environmental governance. *Current Opinion in Environmental Sustainability*, 4(1), 80–87.

Godfray, H. C. J. (2015). The debate over sustainable intensification. *Food Security*, 7(2), 199–208.

Guijt, I. (2008). Rethinking monitoring for collective learning in rural resource management. Doctoral dissertation, Wageningen University, The Netherlands. Retrieved from http://library.wur.nl/WebQuery/wda/lang?dissertatie/nummer=4377 (Accessed 03 April 2014).

GWP (Global Water Partnership)-China, IWA, & IUCN (2014). *Nexus dialogue on water infrastructure solutions building partnerships to optimise infrastructure and technology for water, energy and food security.* Beijing: GWP-China, International Water Association (IWA), and International Union for the Conservation of Nature (IUCN).

Gyawali, D. (2009). Pluralized water policy terrain: Sustainability and integration. Viewpoint in eJournal www.sawasjournal.org.Hyderabad: South Asian Water Studies (SAWAS): www.sawasjournal.org/index.php?option=com_content&view=article&id=48&Itemid=65.

Hall, D., Hirsch, P., and Li, T. M. (2011). *Powers of exclusion: Land dilemmas in Southeast Asia.* Singapore: NUS Press.

Hamilton, J. D. (2009). Causes and consequences of the oil shock of 2007–2008. *Brookings Papers on Economic Activity*, 1(Spring), 215–261.

Hanjra, M. A. and Qureshi, M. E. (2010) 'Global water crisis and future food security in an era of climate change', *Food Policy*, 35.5: 365–377.

Haas, P. M., & Haas, P. M. (1990). *Saving the Mediterranean: The politics of international environmental cooperation.* New York: Columbia University Press.

Hertel, T. W., Burke, M. B., & Lobell, D. B. (2010). The poverty implications of climate-induced crop yield changes by 2030. *Global Environmental Change*, 20(4), 577–585.

Hoff, H. (2011). Understanding the nexus. Background Paper for the Bonn 2011 Conference: The Water, Energy and Food Security Nexus. Stockholm: Stockholm Environment Institute.

Howarth, R. W., Santoro, R., & Ingraffea, A. (2011). Methane and the greenhouse-gas footprint of natural gas from shale formations. *Climatic Change*, 106(4), 679.

Howarth, R. W., Ingraffea, A., & Engelder, T. (2011). Natural gas: Should fracking stop? *Nature*, 477(7364), 271.

Howells, M., Hermann, S., Welsch, M., Bazilian, M., Segerstrom, R., Alfstad, T., Gielen, D., Rogner, H., Fischer, G., Velthuizen, H. van., Wiberg, D., Young, C., Roehrl, R. A., Mueller, A., Steduto P., & Ramma, I. (2013). Integrated analysis of climate change, land-use, energy and water strategies. *Nature Clim. Change*, 3(7), 621–626.

Huff, A. (2015). Understanding relationships between the green economy, resource financialization and conflict. IDS Policy Briefing 95 (July). http://opendocs.ids.ac.uk/opendocs/bitstream/handle/123456789/6461/PB95_AGID573_GreenEconomy_online.pdf;jsessionid=7F3B2DE7D676DB081C52FDB2FEF54063?sequence=1.

IEA. (2010). *World Energy Outlook 2010*. Paris: OECD/ International Energy Agency.

IPCC (2007). *Climate Change 2007: Synthesis Report*. Contribution of Working Groups I, II and III to the Fourth Assessment Report of the Intergovernmental Panel on Climate Change.

Jacobs, M. (2012). Climate policy: deadline 2015. *Nature*, 481(7380), 137–138.

Klare, M. (2011). The global energy crisis deepens: Three energy developments that are changing your life. www.tomdispatch.com/archive/175400/ (accessed on 11 April 2014).

Krchnak, K. M. S., Mark, D., & Duetz, A. (2011). *Putting nature in the nexus: Investing in natural infrastrucutre to advance water-energy-food security*. Bonn: IUCN and The Nature Conservancy.

Krittasudthacheewa, C., Polpanich, O., Bush, A., Srikuta, P., Kemp-Benedict, E., Inmuong, Y., Inmuong, U., Featherston, P., Eagleton, G., Naruchaikusol, S., Pravalpruksul, P., & Krawanchid, D. (2012). *Northeast Thailand futures: A local study of the exploring Mekong Region Futures Project*. Bangkok and Khon Kaen: Stockholm Environment Institute (SEI) and Khon Kaen University.

Lach, D., Rayner, S., & Ingram, H. (2005). Taming the waters: Strategies to domesticate the wicked problems of water resource management. *International Journal of Water Resource Development*, 3, 1–17.

Leach, M., Scoones, I., & Stirling, A. (2010). *Dynamic sustainabilities: Technology, environment, and social justice*. Abingdon: Earthscan.

Leese, M. & Meisch, S. (2015). Securitising Sustainability? Questioning the "Water, Energy and Food-Security Nexus". *Water Alternatives*, 8(1), 695–709.

Lele, U., Klousia-Marquis, M., & Goswami, S. (2013). Good governance for food, water and energy security. *Aquatic Procedia*, 1(0), 44–63.

Lindberg, C. & Leflaive, X. (2015). The water-energy-food-nexus: The imperative of policy coherence for sustainable development. *Coherence for Development, OECD*, 6, December 2015.

Malm, A. (2015). *Fossil capital: The rise of steam-power and the roots of global warming*. London: Verso.

McAfee, K. (2016). The politics of nature in the Anthropocene. In: R. Emmett & T. Lekan (eds). *Whose Anthropocene? Revisiting Dipesh Chakrabarty's 'Four Theses', RCC Perspectives: Transformations in Environment and Society*, 2, 65–72.

Meadows, D. H., Meadows, D. L., Randers, J., & Behrens, W. W. (1972). *The limits to growth: A report for the Club of Rome's project on the predicament of mankind*. New York: Universe Books.

Mehta, L. (ed) (2010). *The limits to scarcity: Contesting the politics of allocation*. London: Earthscan.

Mehta, L., Marshall, F., Stirling, A., Shah, E., Smith, A., Thompson, J., & Movik, S. (2007). Liquid dynamics: Challenges for sustainability in water and sanitation. STEPS Working Paper 6, Brighton: STEPS Centre.

Messer, E. (2009). Rising food prices, social mobilizations, and violence: conceptual issues in understanding and responding to the connections linking hunger and conflict. *NAPA Bulletin*, 32(1), 12–22.

Misselhorn, A., Aggarwal, P., Ericksen, P., Gregory, P., Horn-Phathanothai, L., Ingram, J., & Wiebe, K. (2012). A vision for attaining food security. *Current Opinion in Environmental Sustainability*, 4(1), 7–17.

Molle, F. (2008). Nirvana concepts, storylines and policy models: Insights from the water sector. *Water Alternatives*, 1(1), 131–156.

Moore, J. (2015). *Capitalism in the web of life: Ecology and the accumulation of capital.* London: Verso.

MRC (Mekong River Commission) (2010). Conference summary: MRC International Conference "Transboundary Water Resources Management in a Changing World. 2–3 April 2010, Hua Hin, Thailand. www.mrcmekong.org/news-and-events/speeches/m rc-international-conference-transboundary-water-resources-management-in-a-changing-world/ (Accessed on 22 January 2015).

Mushtaq, S., Maraseni, T. N., Maroulis, J., & Hafeez, M. (2009). Energy and water tradeoffs in enhancing food security: A selective international assessment. *Energy Policy*, 37(9), 3635–3644.

NCVST (2009). *Vulnerability through the eyes of the vulnerable: Climate change induced uncertainties and Nepal's development predicaments.* Institute for Social and Environmental Transition – Nepal (ISET-N) and Institute for Social and Environmental Transition (ISET, Boulder, Colorado) for Nepal Climate Vulnerability Study Team (NCVST).

Nilolova, A., Karazhanova, A., Schneider, N., & Weinberger, K. (2017). Integrated approaches for sustainable development goals planning: The case of Goal 6 on water and sanitation. Bangkok: United Nations.

Ocampo, J. A. (2011). The macroeconomics of the green economy: The transition to a green economy: Benefits, challenges and risks from a sustainable development perspective. Report by a Panel of Experts to Second Preparatory Committee Meeting for United Nations Conference on Sustainable Development.

OECD (2011). *Towards green growth.* Paris: OECD.

Overseas Development Institute (ODI), European Centre for Development Policy Management (ECDPM), & German Development Institute/Deutsches Institut für Entwicklungspolitik (GDI/DIE) (2012). *Confronting scarcity: Managing water, energy and land for inclusive and sustainable growth.* European Report on Development. Brussels: European Union.

Ozment, S., Difrancesco, K., & Gartner, T. (2015). Natural infrastructure in the nexus. Nexus Dialogue Synthesis Papers. Gland, Switzerland.

Pearce, D., Markandya, A., & Barbier, E. (2013). *Blueprint 1: For a green economy.* London: Routledge.

Piesse, J. & Thirtle, C. (2009). Three bubbles and a panic: An explanatory review of recent food commodity price events. *Food Policy* (34), 119–129.

Rasul, G. (2014). Food, water, and energy security in South Asia: A nexus perspective from the Hindu Kush Himalayan region. *Environmental Science & Policy*, 39(0), 35–48.

Reig, P., Luo, T., & Proctor, J. (2014). Global shale gas development: Water availability and business risks. World Resource Institute (WRI). [Online]. Retrieved from: www.wri.org/sites/default/files/wri14_report_shalegas.pdf. [Accessed 26 May 2015].

Rittel, H. & Webber, M. (1973). Dilemmas in a general theory of planning. *Policy Sciences*, 4, 155–169.

Rockström , J., Steffen, W., Noone, K., Persson, Å., Chapin III, F.S., Lambin, E. F., Lenton, T. M., Scheffer, M., Folke, C., Schellnhuber, H. J., & Nykvist, B. (2009). A safe operating space for humanity. *Nature* (461), 472–475.

Rockström, J., Williams, J., Daily, G., Noble, A., Matthews, N., Gordon, L., ... & de Fraiture, C. (2017). Sustainable intensification of agriculture for human prosperity and global sustainability. *Ambio*, 46(1), 4–17.

Rosegrant, M. W., Ringler, C., Zhu, T., Tokgoz, S., & Sabbagh, P. (2012). Water and food security in the Greater Mekong Subregion: Outlook to 2030/2050. International Conference on GMS 2020: Balancing economic growth and environmental sustainability, focusing on food-water-energy nexus. Bangkok, Thailand: Asian Development Bank (ADB).

Royal Society (2008). *Sustainable biofuels: Prospects and challenges.* London: Royal Society.

Ruel, M. T., Garrett, J. L., Hawkes, C., & Cohen, M. J. (2009). The food, fuel, and financial crises affect the urban and rural poor disproportionately: A review of the evidence. *J Nutr*, 140(1), 170S–176S.

SABMiller / WWF (2014). The water-food-energy nexus: Insights into resilient development. http://assets.wwf.org.uk/downloads/sab03_01_sab_wwf_project_nexus_final.pdf (accessed 1 October 2018).

Salam, P. A., Shrestha, S., Pandey, V. P., & Anal, A. K. (eds) (2017). *Water-energy-food nexus: Principles and practices.* New Jersey: John Wiley & Sons.

Schmidhuber, J. & Tubiello, F. N. (2007). Global food security under climate change. *Proceedings of the National Academy of Sciences*, 104(50), 19703–19708.

Scoones, I., Smalley, R., Hall, R., & Tsikata, D. (2014). Narratives of scarcity: Understanding the "global resource grab" (Working Paper 76). Cape Town: Institute for Poverty, Land and Agrarian Studies (PLASS).

Schlör, H., Venghaus, S., Fischer, W., Märker, C., & Hake, J. F. (2018). Deliberations about a perfect storm: The meaning of justice for food energy water-nexus (FEW-Nexus). *Journal of Environmental Management*, 220, 16–29.

Smajgl, A. & Ward, J. (eds) (2013). *The water-food-energy nexus in the Mekong region: Assessing development strategies considering cross-sectoral and transboundary impacts.* New York, Heidelberg, Dordrecht, and London: Springer.

Steffen, W., Richardson, K., Rockström, J., Cornell, S.E., Fetzer, I., Bennett, E.M., Biggs, R., Carpenter, S.R., De Vries, W., De Wit, C.A., & Folke, C. (2015). Planetary boundaries: Guiding human development on a changing planet. *Science*, 347 (6223), 1259855.

Stott, P. (1998). Biogeography and ecology in crisis: the urgent need for a new metalanguage. *J Biogeogr*, 25, 1–2.

Thapan, A. (2012). The future of water in the GMS: Is it history? International Conference on GMS 2020. Balancing economic growth and environmental sustainability, focusing on food-water-energy nexus. Bangkok, Thailand: Asian Development Bank (ADB).

Thompson, M., Grendstad, G., & Selle, P. (eds) (1999). *Cultural theory as political science.* London: Routledge.

Thompson, M. (2008). *Organising and disorganising: A dynamic and non-linear theory of institutional emergence and its implications.* Axminster, UK: Triarchy Press.

UNESCAP (2013). Water, food and energy security in Asia and the Pacific. A position paper commissioned by the United Nations Economic and Social Commission for Asia and the Pacific, UNESCAP.

Unmüßig, B., Sachs, W., & Fatheuer, T. (2012). *A critique of the green economy: Toward social and environmental equity.* Berlin: Publications-Heinrich Böll Foundation.

Verweij, M. & Thompson, M. (eds) (2006). *Clumsy solutions for a complex world.* Basingstoke, UK: Palgrave/Macmillan.

Von Grebmer, K., Fritschel, H., Nestorova, B., Olofinbiyi, T., Pandya-Lorch, R., & Yohannes, Y. (2008). *Global Hunger Index: The challenge of hunger.* Washington DC: IFPRI.

Wales, A. (2014). Making sustainable beer *Nature Climate Change*, 4(5), 316–318.

Wales, A. & Winston, A. (2012). *Ecosystem economics: Navigating the water-food-energy nexus.* London: The Guardian Sustainable Business Partner Zone, www.theguardian.co.uk (accessed on 13 April 2014).

Weitz, N., Nilsson, M., & Davis, M. (2014). A nexus approach to the post-2015 agenda: Formulating integrated water, energy, and food SDGs. *SAIS Rev. Int. Aff.,* 34(2), 37–50.

World Business Council for Sustainable Development (2014). *Water, food and energy nexus challenges.* Geneva: WBCSD.

WEF (2009). *Thirsty energy: Water and energy in the 21st century.* World Economic Forum in partnership with Cambridge Energy Research Associates.

WEF (2011a). *Water security: The water-food-energy-climate nexus.* Washington: Island Press.

WEF (2011b). *Global risks 2011* (Sixth Edition). Cologny/Geneva: WEF.

WEF (2014). The water-energy nexus: Strategic considerations for energy policy-makers, global agenda council on energy security. (May 2014). Available at: www3.weforum.org/docs/GAC/2014/WEF_GAC_EnergySecurity_WaterEnergyNexus_Paper_2014.pdf.

Working Group on Green Economy (2012). *Is the "green economy" a new Washington consensus?*Montreal: Global Research Available at www.globalresearch.ca/is-the-green-economy-a-new-washingtonconsensus/29462. (Accessed on 16 January 2014).

WWF (2014). *Luc Hoffmann Institute Fellow (The Mekong Nexus Project) at Australian National University (ANU).* wwf.panda.org/who_we_are/jobs/?224399/luc-hoffmann-institute-fellow-the-mekong-nexus-project-at-australian-national-university-anu. (Accessed on 20 January 2015).

3

INTEGRATION FOR WHOM?

Learning from the past

The nexus has gained momentum at the expense of Integrated Water Resources Management (IWRM), which has, according to Bird (2012: 397), "tended to stay within the domains of the water, agriculture and environment professionals and not had much traction with energy sector professionals". However, for some water specialists, the nexus may be viewed as old wine in new bottles. The presentation of hard statistics about water consumption in food and energy and the proposal for nexus approaches, may give a veneer of newness to global policy makers, and yet looking in more detail at the discourse of the nexus to date, there is far less clarity on what a new common integrative approach might look like beyond the existing water-centric paradigm of IWRM (see Mueller 2015 and Wichelns 2017). Indeed, seen this way the nexus is based on the same idea of IWRM where there is a need for cross-sectoral integration to accommodate the need for water for people, water for food, water for nature, and water for industry.

One of the major debates in IWRM beyond what the meaning of integration was, was integration for whom, and looking back at debates on IWRM helps to shed some light on the nexus for whom. IWRM's most commonly used definition is a "process which promotes the coordinated development and management of water, land and related resources, in order to maximise the resultant economic and social welfare in an equitable manner without compromising the sustainability of vital ecosystems" (GWP 2000). IWRM has used the language of equity, and contained the following questions:

- What is precisely meant by equitable?
- How will this be determined operationally?
- Who will decide what is equitable, for whom, and from what perspectives and under what conditions?

In contrast to IWRM's focus on equity, the globally-dominant interpretation of the nexus rather uses the language of human security. However, the equitable dimensions of IWRM or the human security aspects of the nexus are not really addressed and debated at the global level, which assumes that these development benefits will trickle down. However, as history tells us, it is unlikely to be the case as will be discussed in this chapter.

3.1. Sectoral and spatial integration of resource management: Still looking for successes

Despite the advantages a water–food–energy nexus approach brings by increasing integration across research, planning, and management, the concept also faces criticism. It is often argued that the integrated approach offered by the nexus concept is not new. The interactions between water, energy, and food systems have been known and studied for many years by scientists and policy analysts alike; the need for greater integration of research and policy discourse has been expressed since the late 1940s (Wichelns 2017). Previous attempts to improve integration across sectors and spatial scales, through, for example, IWRM and Integrated Natural Resource Management, have to date achieved relatively little in terms of integration and policy coherence. The nexus concept has yet to have any greater success than these other existing integrated approaches (Smajgl, Ward, & Pluschke 2016; Wichelns 2017).

Despite the apparent "newness" of the nexus concept, elements of this approach are historically evident. Molle (2009) shows how the integration of water resources management at river basin scales dates back many decades, if not centuries, and involves several semi-distinct paradigmatic changes. Different experiments around integrated water management, one could argue, go back to as early as the beginning of the 20th century (if not before), notably in the US and Europe (Mitchell 1990; Rahaman and Varis 2005; Teclaff 1967).

The Tennessee Valley Authority (TVA), which was established in 1933 in the US, was an early example of integrated resource management in practice, as it was set up as a river basin organisation to facilitate multi-purpose management to deal with water supply, pollution, navigation, flood management, and conservation. The TVA effort contained many elements of today's principles and values of integration: comprehensive planning of natural resource utilisation combined with economic, social, and even environmental objectives (Cherlet 2012; Mukhtarov 2009; Snellen & Schrevel 2004).

Such integrated approaches then became the blueprint for developing countries "as large-scale water engineering projects became a means to drive national development strategies" (Gain, Rouillard, & Benson 2013: 12). Beginning in the 1940s, this notion spread as "TVA-like river-basin development plans mushroomed all over the world" (Molle 2009: 489), particularly in Asia, Africa, and South America. This approach was overtly engineering-based and development oriented, with little real consideration of other values (Gain, Rouillard, & Benson 2013). Thereafter, the era between the late 1970s and early 1990s was characterised by "a certain demise and loss of appeal of the river-basin concept" (Molle 2009: 490), leading to new thinking. Attempts were made to regulate emerging point source water pollution problems through national and European level legislation, for example the US Clean Water Act amendments of 1972 and European Economic Community water directives (Benson, Jordan, & Smith 2013). But the failure of "command and control" centralised approaches to deal with diffuse water pollution in particular led to demands for more integrated management at the river basin scale. Locally collaborative "watershed" management began to emerge during the 1980s in the US due to a variety of factors, including dissatisfaction with federal controls (Sabatier, Weible, & Ficker 2005). There was then a growing perception amongst water professionals globally that a new paradigm was required to better reflect the multidimensional nature of water management (Biswas 2008a). By the early 1990s, these views had been formalised into a new concept – IWRM.

3.2. IWRM

Since the early 1990s, IWRM[1] became the rallying point for international water policy (Conca 2006), leading scholars such as Jeffrey and Gearey (2006: 2) to argue that it has become the orthodoxy in water resources management. The concept gained momentum at the International Conference on Water and the Environment (ICWE) in Dublin, in January 1992, at which time the "Dublin Principles" were proposed. Since then it has been the flagship project of supranational bodies such as the Global Water Partnership (GWP) (Finger & Allouche 2002; GWP 2000). It is central to the European Water Framework Directive (European Union 2000; Kaika 2003). It has also been actively promoted by multilateral and regional development banks, as well as by bilateral donor agencies. The promotion of IWRM by these global players has led to a quasi-global industry around IWRM manifesting itself in various forms such as master's degrees and short courses, annual symposia such as WaterNet in southern Africa, IWRM toolkits, and manuals (Mehta et al 2016). The World Summit on Sustainable Development in 2002 called for all countries to draft IWRM and water efficiency strategies by the end of 2005. According to Cherlet (2012), over 80 per cent of countries worldwide now have the IWRM principles in their water laws and two thirds have developed a national IWRM plan. Hassing (2009) reported the findings of a survey done for the 4th World Water Forum in Mexico, which showed that about three-quarters of the 95 countries for which responses were available used IWRM terminology in at least one policy or law, the vast majority of which were created after 2002. The 2012 World Water Development Report reports that more than one hundred countries have implemented IWRM (WWAP 2012). All these facts and figures show how the concept of IWRM is very popular among governments and international organisations, and within the water expert community (Allouche 2016), despite questions about the efficacy of its actual practice.

Demands for new approaches have, however, emerged in the 2000s. The introduction of resilience to global water governance led to calls for adaptive, experimental forms of management to augment IWRM in recognition of complexity and connectivity in complex human-hydrological systems (e.g. Galaz 2007). Water experts pushed for transitions towards adaptive management as a way to retain the goal of integrated management but with deeper appreciation of existing social-ecological dynamics at stake in sustainable development. Adaptive water management (AWM) can be traced back to decision-making practices that emerged in the US and Australia during the 1990s. Adaptive management (AM) refers a systematic process for continually improving management practices by learning new experiences and insights from the outcomes of implemented policies (Engle et al 2011; Feldman 2007; Holling 1978; Pahl-Wostl 2007; Pahl-Wostl et al 2007). AM involves the approach "learning by doing" as feedback mechanisms from the environment (biophysical and/or social) are monitored and analysed to then (re)shape policy, followed by further systematic experimentation and learning, in a never-ending cycle (Allen et al 2011; Berkes, Colding, & Folke 2002; Walters and Holling 1990). Based on the principles of AM, AWM features the use of stakeholder input and knowledge generation, objectives setting, management planning, monitoring implementation, and incremental plan adjustment in the face of uncertainty (Engle et al 2011; Huitema et al 2009; Pahl-Wostl 2007; Pahl-Wostl et al 2008). The potential of AM was the retention of IWRM's holistic goals in a register that connected the "blue water" of industrial inputs to the "green water" effects of land-use change, energy production, and climate change (Falkenmark 2004).

The history of integrated water management has therefore undergone several inter-connected shifts, resulting in the current IWRM paradigm. And now, finally, is there IWRM fatigue? Are we witnessing the death of IWRM? Some have argued that IWRM is now slowly becoming replaced by other emerging concepts such as water security or the nexus (Allouche, Middleton, & Gyawali 2015). Indeed, the concept may become irrelevant if one focuses on new challenges (such as adaption and climate change or the water–food–energy crises). The importance of new donors such as the BRICS countries may also challenge the concept of IWRM, as it may not fit with their key priorities around infrastructural development. The answer remains to be seen, but the GWP has been quite strategic in adapting its approach to IWRM with other challenges and concepts, aligning the IWRM concept with climate change discourses and water security.

One of the major debates in IWRM is the meaning of integration, and integration for whom. Despite the very important aims of holistic and integrated approaches in water management, the concept of IWRM remains ideal-typical, a nirvana concept (Molle 2008), and is beset by contradictions, which makes it very difficult to operationalise and implement on the ground. This complexity is reflected just by looking at understanding the notion of integration. As Bolding, Molinga, and Zwarteveen (2000) have argued, there are at least four possible meanings of integration, including: (1) the integration of different uses of water (e. g. drinking, irrigation, ecological functions, manufacture, etc); (2) the integration of analytical perspectives and the fact that the organisation of knowledge production tends to be along disciplinary and sectoral lines, making it a challenge to integrate different disciplinary and sectoral approaches; (3) the integration of the different institutions responsible for water resources development and management and the need to break down sectoral compartmentalisation between different ministries (e. g. irrigation, rural water supply, forestry, land, and so on); these raise the more general issue of how to organise stakeholder involvement in water and natural resources policy-making, planning, development and management; (4) finally, water management as integrated with ecosystem services, human health, ecological sustainability, economic growth, poverty alleviation, gender equality, employment, and other aspects of human development.

IWRM has been critiqued from many different perspectives (Allan 2003; Biswas 2008b; Gyawali & Dixit 1999; Jeffrey & Greary 2006; Moench et al 2003) but what is common to all is the failure of the IWRM approach to answer two fundamental questions which tellingly remains as residues in the "nexus" debate as well:

- Who does the integrating? (agency) and
- How? (process)

IWRM was seen and presented as a break from the technocratic, supply-driven, and construction-oriented paradigm of the past, to signal a new era in which economic, social and environmental concerns are addressed simultaneously and in their mutual interactions. However, Parés (2011) argues that new water governance arrangements are foreclosing the possibility to change the established system, leading to increasing managerialism and consensual politics, a sentiment that is echoed in Swyngedouw's work on "post-democratisation", where he explores the dynamics of de-politisation and the erosion of democracy and shrinking public spheres (Swyngedouw 2011). A post-democratic form of governance reduces value-laden political issues to technicalities that can best be dealt with by experts.

Integration is a political process about allocations priorities and trade-offs among different water uses and sectors (Saravanan, McDonald, & Mollinga 2009) and we need to ask who is doing the integrating, whose interests are being represented, and how contested interests should be dealt with (Merrey et al 2005). These political economy aspects are often ignored because integration implicitly draws on a normative logic of Habermasian communicative rationality, where different members seek to reach a common understanding and cooperative actions by consensus – and an assumed equality that is critiqued by Fraser (1990) – rather than strategic action strictly pursing their own goals (Saravanan, McDonald, & Mollinga 2009).

As a growing body of political ecology and water justice studies have shown, IWRM, despite its emphasis on the three Es – equity, efficiency, and environmental sustainability – is often used to hide or sanction processes of dispossession and accumulation of water, processes that are far from democratic or participatory (Allan 2006; Molle, Mollinga, & Wester 2009). IWRM does not explicitly focus on poverty reduction issues and wider development concerns (see Swatuk 2005). IWRM, by its focus on "second generation issues" such as demand management and water re-allocation, could be harmful in a range of African contexts, for example, where water resources development is often more necessary than management measures and where agriculture contributes to poverty reduction and livelihood security (Merrey et al 2005).

Lessons from IWRM in Southeast Asia show us similar challenges (GWP 2011; Molle 2007). Reflecting on intergovernmental attempts to implement an IWRM Basin Development Plan through the Mekong River Commission for the lower Mekong Basin, Hansson, Hellberg, and Öjendal (2012) argue that greater attention must be paid to power asymmetries and politics in regional water politics if transboundary water management is to be sustainable and inclusive (see also Cooper 2012).

"Efficient" water allocations are justified by a scarcity discourse, but not everyone is equally threatened by water scarcity (Bakker 2004; Loris 2012), and accumulation by some often goes hand in hand with deepening scarcity as experienced by others (Zwarteveen & Boelens 2014). IWRM discourses and the allocations they sanction create rankings of water uses and users on the basis of specific calculations of efficiency, with the most efficient uses and users being awarded the premium of modernity and water citizenship. "Modern" users – such as large-scale commercial enterprises, agribusiness firms, private drinking-water companies, and mining and hydropower conglomerates – thus become the example to be followed, representing the ideals of water use efficiency and water market rationality that science preaches (Boelens & Vos 2012). In contrast, people who use traditional irrigation systems for growing their own food crops come to be seen as "backward". For water scarcity problems to be overcome, they either need to disappear or they need to correct their water misbehaviour to join "progress" and "development" (Castro 2007; Vera & Zwarteveen 2008).

While IWRM achieved the status of a powerful mantra incanted at various policy levels from the national to the global, it remains un-operationalised or even un-operationalisable at the level of governance where it matters. Even when codified in formal policy documents such as national irrigation or water policies, the official water management agencies have continued with their business-as-usual construction contracts for cement and earth-moving with little thought given to water, farmers, or their crops, and even less to the energy for processing, storage, and transport (Gyawali 2013). It has led in activist circles to the cynical conclusion that in reality, IWRM was dressed to please but bereft of genuine integration of myriad complex concerns. Indeed, what remained unanswered were issues such as: whose interests should be reflected in the integrating process and how should such a process be governed to ensure that the interests of all stakeholders, especially those of the poor and the marginalised, are adequately reflected; how should disputes be resolved; or, when can some water management issues be addressed in isolation (waste disposal from treatment plants) while others (water allocation in conditions of scarcity) canno;, which issues need integrated treatment and which issues are more amenable to "silo" approaches?

3.3. Nexus and IWRM

Market solutions to resource scarcity and attempts at integration to multiple demands as two key policy solutions are not very novel. Previous concepts such as

IWRM were also promoting integrated approaches and water as an economic good. And, as discussed above, the idea of integration at the river basin level is a very old idea (Molle 2009). Thus, the nexus may be seen as a repetition of the debate of the early nineties with the Dublin Principles and the emergence of IWRM. The following quote by the World Economic Forum (WEF) shows this continuity by emphasising how water is an economic and social good:

> Water lies at the heart of a nexus of social, economic and political issues – agriculture, energy, cities, trade, finance, national security and human lives, rich and poor, water is not only an indispensable ingredient for human life, seen by many as a right, but also indisputably an economic and social good unlike any other. It is a commodity in its own right ... but it is also a crucial connector between humans, our environment and all aspects of our economic system.
>
> *WEF 2011: 3*

Benson, Gain, and Rouillard (2015) argue that the nexus presents some novel elements compared to the IWRM paradigm, particularly in terms of holistically integrating different policy sectors, encouraging business involvement, promoting economically rational decision-making, and privileging water securitisation in the pursuit of sustainable development (Table 3.1). According to Bach et al (2012) the critical difference is that the nexus is a multi-centric concept that treats the different sectors – water, energy, food, and climate security – as equally important as opposed to IWRM, which is water-centric. However, recent research has shown that most of the more than half (55%) of the current English-language peer reviewed articles published in 2016 showed some preference for one sector. Preference was most commonly given to the water sector (21% of the total), then

TABLE 3.1 Key features on the water security nexus and IWRM

	Nexus	*IWRM*
Integration	Integrating water, energy and food policy objectives	Integrating water with other policy objectives
Optimal governance	Integrated policy solutions Multi-tiered institutions	"Good governance" principles
Scale	Multiple scales	River basin scale
Participation	Public–private partnerships – multistakeholder platforms for increasing stakeholder collaboration	Stakeholder involvement in decision-making Multiple actors, including women
Resource use	Economically rational decision-making Cost recovery	Efficient allocations Cost recovery Equitable access
Sustainable development	Securitisation of resources	Demand management

Source: Benson, Gain, & Rouillard 2015

followed by energy and food sectors (8% and 7%, respectively) (Albrecht, Crootof, & Scott 2018). The IWRM and nexus approaches are very closely related. The ultimate objectives of both are to promote better resource use to allow societies to develop in ways which are sustainable – environmentally, socially, and economically. A general overview would certainly suggest several similarities but also differences around their normative assumptions on policy integration, optimal governance, scales, stakeholder participation, resource use, and sustainable development (Benson, Gain, & Rouillard 2015).

However, there is variance between IWRM and the nexus in the optimal scale at which interaction is anticipated. As promoted by international agencies and national governments, IWRM is overtly premised on institutional forms at the regional, river basin (hydrological) scale while also providing an overarching centralised approach for national policy (Rouillard, Benson, & Gain 2014). In other words, in terms of implementation, it is a national and river basin scale approach. These norms now underpin national water governance strategies in many countries worldwide (see Gain, Rouillard, & Benson 2013). In contrast, the nexus, in its original WEF conception, includes a broader set of macro- or meso-scale norms for integrating policy sectors (i.e. "policy coherence") between different levels, but provides limited guidance on how this should, normatively, occur in practice. That said, different studies have examined the nexus at specific scales, including the river basin, demonstrating its multi-level, holistic nature (Lawford et al 2013; Newell, Marsh, & Sharma 2011; Pittock 2011; Opperman et al 2011; Stillwell et al 2011).

IWRM was criticized for being too "water-centric", whereas the nexus remained committed to coordinated, sustainable development, but focused on governing connections across interconnected sites affecting water, such as energy and food production and climate change (Benson, Gain, & Rouillard 2015; Hussey & Pittock 2012). Since, in a globalised world, interconnected concerns extend beyond the watershed, the nexus offered a framework to connect watersheds to the institutional, political, and economic scales that govern global supply-chains of water, food, and energy – from Spain, India, China, and Mexico to the United States (Hardy, Alberto, & Juana 2012; Malik 2002; Scott 2011; Scott et al 2011; Shah et al 2003; Wang et al 2012).

More fundamentally perhaps, the nexus reveals a broader shift from state-oriented development models to financialised approaches of water development and sustainability (Conca 2015; Schmidt & Matthews 2018). As put by Schmidt and Mathews (2018), financial technologies shape the kinds of connections that are to be governed and secured as development agendas are re-scaled from state-centred modes of industrial production to financialised systems of accumulation, by crafting a political and conceptual vocabulary around the nexus that could shift state-oriented, industrial development to financialised approaches to sustainability. In this regard, the nexus facilitates the shift from "state to system" by shifting both the objects of governance and the techniques used to disclose connections among them. The opportunities to open up pluralistic understandings of the world are being foreclosed upon by financial languages of risk and resilience as rapidly as financialised approaches to sustainability identify opportunities to accumulate profits that outpace those derived from the traditional aims of sustainable industrialism.

The social justice critique of the nexus echoes very much the same debates with IWRM. Williams et al (2014) argue the nexus is not a significant departure from previous sustainable development programs, such as IWRM, but rather retains neoliberal logics where capitalist modes of production create and shape spaces for accumulation. Both Leese and Meisch (2015) and Biggs et al (2015) also consider that the emphasis on security and securitisation marginalise distributive and livelihood concerns.

3.4. Integration for the benefit of which system? Integration for whom?

Though the nexus evokes an "integrative imaginary" (Cairns & Krzywoszynska 2016) as an integration across various sectors, the boundaries of the nexus are still very much disputed (Pittock, Hussey, & McGlennon 2013; Rees 2013; Srivastava & Mehta 2014). If we understand nexuses to be interlinkages across different kinds of resources – water, energy, food, water, climate – then we are essentially referring to not only interdependent systems (as emphasised by the nexus thinking) but inherently complex systems. Bazilian et al (2011) reveal the complexity of this interconnectedness when they identify analytical and policy-making entry points. They argue that:

> If a water perspective is adopted, then food and energy systems are users of the resource (see e.g., Hellegers and Zilberman 2008); from a food perspective energy and water are inputs (see e.g., Mushtaq et al. 2009; UN-DESA 2011; Khan and Hanjra 2009); from an energy perspective, water as well as bio-resources (e.g., biomass in form of energy crops) are generally an input or resource requirement and food is generally the output. Food and water supply as well as wastewater treatment require significant amounts of energy. Of course, areas such as food-as-fuels (i.e., biofuels) tend to blur these descriptions …'
>
> *Bazilian et al 2011: 7897*

This becomes more complex as ecological, social, technological, economic, and political systems that govern these systems interact within themselves in the backdrop of climate change (see Leach, Scoones, & Stirling 2010). For example, the hydrological is a highly dynamic system socially, economically, and technologically, and will become even more dynamic under the conditions of climate change (Mehta et al 2007). Given that food, water, and energy sectors often exist in silos, integration may be challenging to put into practice; that water, land, and energy have different governing regimes will make nexus governance even more difficult. It is not surprising to see the emergence of parallel concepts, which rather emphasise the two sub-nexuses, the water food trade sub-nexus and the energy–climate change sub-nexus (Allan et al 2015). In fact, this was acknowledged in the

WEF (2008) global risk reports where two competing constellations of a water–food–trade nexus and an energy–climate nexus shaped the WEF's approach to connecting global economic risks with those to water, food, energy, and climate (Allan, Keulertz, & Woertz 2015). By 2009, however, water assumed central place in a single nexus because of its fundamental role across energy and food production (see WEF 2009). These two sub-nexuses are separate and very different in character and operate in very different global regimes, any changes need to come from an international political economy perspective, a point we will come back to later, rather than a "prescribed optimal scientific rational" one.

For many, water security remains central to the concept of the nexus, especially in the sense that it is mostly the international water policy sector which has taken up the language of the nexus (compared to those focusing on energy and food). Though water and energy are closely linked in the production phase, water security is prioritised in the nexus debates; in short, food and energy security can only be achieved through water security. Climate change appears on a relative periphery of this debate and has not been the primary driver for change.

The dominant approach to nexus thinking is through socio-ecological systems thinking that seek to understand trade-offs and synergies, increase efficiency, and improve governance between food, water, and energy systems (e.g. Hoff 2011; Smajgl & Ward 2013). The nexus seeks to integrate sectors through making visible and thereby aiming to address externalities that link sectors together. It therefore raises a need to discuss trade-offs and the decisions that this entails; the governance of these decisions, namely who takes decisions and for whom, is clearly important and yet less discussed in nexus policy circles. If the nexus approach is to support its commonly stated aspirations for sustainable development and poverty reduction, then it should engage more directly in identifying winners and losers in nexused natural resource decision-making, the inevitable politics involved, and ultimately with the issue of justice. To date, nexus framings that adopt a systems perspective, whilst broadly calling for "good governance", are yet to seriously meet this challenge (Allouche, Middleton, & Gyawali 2014; Foran 2015; Lele, Klousia-Marquis, & Goswami 2013). We return to the question of justice in the nexus in Chapter 6.

One way, to date, that resource managers have sought to address the integration question is to focus on a particular policy issue linked to the nexus, for example the argument of having more large dams and storage devices as a way to ensure water security and manage the trade-offs between water, energy, and food while at the same time addressing the imperatives of dealing with climate change for clean energy (Hoff et al 2012). Sean Cleary, in the WEF's report, highlights the vicious circle of energy, water, and climate change, arguing that "we need more energy for more development but the current processes of energy production put pressure on water availability which has an impact on climate; climate change variability affects water availability patterns and that, in turn, affects energy production" (WEF 2009a). For those more centrally concerned with climate change, water storage from large dams kills two birds (water–food–energy as well as climate change impacts) with one simple nexus stone. Lall (WEF 2011) argues that in the backdrop of climate change and climate variability, the key question that global society faces is:

> how should our water best be stored and which stores should be used to minimize risks due to long term climate variability and change. Storage guarantees reliability in water supply, which in turn means food security, electricity generation and industrial growth?

Lack of water storage infrastructure is predicted by some as a serious impediment to economic growth in developing countries (Grey and Sadoff 2007). Climate change also has implications for existing infrastructure – for example, dams designed in the past without accounting for the increasing variability of climate change are now increasingly at risk (i.e. 100 year floods may be more severe, meaning that infrastructure is under-designed).

The move from IWRM to the nexus would appear to be a giant leap into interdisciplinarity (see Chapter 4) and inter-sectoral complexity of natural resources management such as water, global atmosphere (with climate change), forests, or oceans. While IWRM was asking of (water) sector managers to be more broad-based and accommodating in their approach, the nexus is asking managers to think like managers of other resources, imbibe their concerns and then only decide measures in their own sectors. It also nudges different sector managers to desire and listen to a supra-sectoral integrating or coordinating body/platform where such concerns can be better appreciated than within one's silos. We now turn to the challenges of integration and interdisciplinarity, and in particular the role of knowledge production.

Note

1 Another definition by USAID says: "IWRM is a participatory planning and implementation process, based on sound science, which brings together stakeholders to determine how to meet society's long–term needs for water and coastal resource while maintaining essential ecological services and economic benefits. (…) IWRM helps to protect the world's environment, foster economic growth and sustainable agricultural development, promote democratic governance, and improve human health" (cited in Xie 2006).

References

Albrecht, T. R., Crootof, A., & Scott, C. A. (2018). The water-energy-food-nexus: A systematic review of methods for nexus assessment. *Environ. Res. Lett*, 13, in press.

Allan, T. (J. A.) (2003). IWRM/IWRAM: A new sanctioned discourse? SOAS/London: Occasional Paper 50.

Allan, T. T. (2006). IWRM: The new sanctioned discourse? In P. Mollinga, A. Dixit, & K. Athukorala (eds). *Integrated water resources management* (pp. 38–63). New Delhi: Sage.

Allan, T., Keulertz, M., & Woertz, E. (2015). The water-food-energy nexus: an introduction to nexus concepts and some conceptual and operational problems. *International Journal of Water Resources Development*, 31(3), 301–311.

Allen, C. R., Fontaine, J. J., Pope, K. L., & Garmestani, A.S. (2011). Adaptive management for a turbulent future. *Journal of Environmental Management*, 92, 1339–1345.

Allouche, J. (2016). The birth and spread of IWRM: A case study of global policy diffusion and translation. *Water Alternatives, 9*(3), 412.

Allouche, J., Middleton, C., & Gyawali, D. (2014). Nexus nirvana or nexus nullity? A dynamic approach to security and sustainability in the water-energy-food nexus. STEPS Working Paper 63, Brighton: STEPS Center. Downloadable at: http://steps-centre.org/wp-content/uploads/Water-and-the-Nexus.pdf.

Allouche, J., Middleton, C., & Gyawali, D. (2015) Technical veil, hidden politics: Interrogating the power linkages behind the nexus. *Water Alternatives*, 8(1): 610–626. Downloadable at: www.water-alternatives.org/index.php/alldoc/articles/vol8/v8issue1/277-a8-1-1/file.

Bach, H., Bird, J., Clausen, T. J., Jensen, K. M., Lange, R. B., Taylor, R., Viriyasakultorn, V., & Wolf, A. (2012). *Transboundary river basin management: Addressing water, energy and food security.* Lao, PDR: Mekong River Commission.

Bakker, K. (2004). *An uncooperative commodity: Privatizing water in England and wales.* Oxford: Oxford University Press.

Bazilian, M., Rogner, H., Howells, M., Hermann, S., Arent, D., Gielen, D., & Yumkella, K. K. (2011). Considering the energy, water and food nexus: towards an integrated modelling approach. *Energy Policy*, 39(12): 7896–7906.

Benson, D., Jordan, A.J., & Smith, L. (2013). Is environmental management really more collaborative? A comparative analysis of putative 'paradigm shifts' in Europe, Australia and the USA. *Environment and Planning A*, 30(6).

Benson, D., Gain, A. K., & Rouillard, J. J. (2015). Water governance in a comparative perspective: From IWRM to a "nexus" approach? *Water Alternatives*, 8(1), 756–773.

Berkes, F. L., Colding, J., & Folke, C. (eds) (2002). *Navigating social-ecological systems: Building resilience for complexity and change.* Cambridge, UK: Cambridge University Press.

Boelens, R. & Vos, J. (2012). The danger of naturalizing water policy concepts: Water productivity and efficiency discourses from field irrigation to virtual water trade. *Agric Water Manag*, 108, 16–26.

Bolding, A., Mollinga, P.P., & Zwarteveen, M. (2000). Interdisciplinarity in research on integrated water resource management: Pitfalls and challenges. Paper presented at the Unesco-Wotro international working conference on 'Water for Society', Delft, the Netherlands, 8–10 November.

Biggs, E., Bruce, E., Boruff, B. et al. (2015). Sustainable development and the water-energy-food nexus: A perspective on livelihoods. *Environmental Science & Policy*, 54, 389–397.

Bird, J. (2012). Water resources management in the Greater Mekong subregion: Linkages to hydropower planning for a sustainable future. GMS 2020 Conference: Balancing Economic Growth and Environmental Sustainability. Bangkok: Asian Development Bank, 377–389.

Biswas, A. K. (2008a). Integrated Water Resources Management: Is It Working? *Water Resources Development*, 24(1), 5–22.

Biswas, A. K. (2008b). Current Directions: Integrated Water Resources Management: A Second Look. *Water International*, 33(3), 274–278.

Cairns, R. & Krzywoszynska, A. (2016). Anatomy of a buzzword: The emergence of "the water-energy-food nexus" in UK natural resource debates. *Environmental Science & Policy*, 64, 164–170.

Castro, J. E. (2007). Poverty and citizenship: Sociological perspectives on water services and public–private participation. *Geoforum*, 38(5), 756–771.

Cherlet, J. (2012). Tracing the emergence and deployment of the "Integrated Water Resources Management" paradigm. Paper presented at the 12th EASA Biennial Conference, Nanterre, 10–13 July 2012, Unpublished document.

Conca, K. (2006). *Governing water: Contentious transnational politics and global institution building.* Cambridge, MA: MIT Press.

Conca, K. (2015). Which risks get managed? Addressing climate effects in the context of evolving water-governance institutions. *Water Alternatives*, 8(3), 301–316.

Cooper, R. (2012). The potential of MRC to pursue IWRM in the Mekong: Trade-offs and public participation. In J. Öjendal, S. Hansson, & S. Hellberg (eds). *Politics and development in a transboundary watershed: The case of the Lower Mekong Basin* (pp. 61–82). Dordrecht, Heidelberg and London, New York: Springer.

Engle, N. L., Johns, O. R., Lemos, M., & Nelson, D.R. (2011). Integrated and adaptive management of water resources: tensions, legacies, and the next best thing. *Ecology and Society*, 16(1), 19. www.ecologyandsociety.org/vol16/iss1/art19/.

European Union (2000). Directive 2000/60/EC of the European Parliament and of the Council establishing a framework for the community action in the field of water policy. http://ec.europa.eu/environment/water/water-framework/index_en.html (accessed September 2016).

Falkenmark, M. (2004). Towards integrated catchment management: Opening the paradigm locks between hydrology, ecology and policy-making. *Int. J. Water Resour. Dev.*, 20(3), 275–281.

Feldman, D. (2007). *Water policy for sustainable development.* Baltimore, MD: John Hopkins University Press.

Finger, M. & Allouche, J. (2002). *Water privatization: Trans-national corporations and the re-regulation of the water industry.* London: Spon Press.

Foran, T. (2015). Node and regime: Interdisciplinary analysis of water-energy-food nexus in the Mekong region. *Water Alternatives*, 8(1), 655–674.

Fraser, N. (1990). Rethinking the public sphere: A contribution to the critique of actually existing democracy. *Social Text*, 25/26, 56–80.

Gain, A. K., Rouillard, J. J., & Benson, D. (2013). Can integrated water resources management increase adaptive capacity to climate change adaptation? A critical review. *Journal of Water Resource and Protection*, 5(4A), 11–20.

Galaz, V. (2007). Water governance, resilience and global environmental change – a re-assessment of integrated water resources management (IWRM). *Water Sci Technol.*, 56(4), 1–9.

Grey, D. & Sadoff, C.W. (2007). "Sink or swim? Water security for growth and development. *Water Policy*, 9, 545–571.

GWP (Global Water Partnership) (2000). *IWRM components.* Stockholm: GWP. www.gwp.org/en/The-Challenge/What-is-IWRM/IWRM-Principles/ (accessed 2 May 2015).

GWP (Global Water Partnership) (2010). Integrated water resource management. Technical Advisory Committee Background Paper No. 4. Stockholm, Sweden: Global Water Partnership.

GWP. (2011). *Southeast Asia – Evaluation of the status of IWRM: Implementation in Southeast Asia 2000–2010 in respect to policy, legal and institutional aspects.* Bangkok: Global Water Partnership (GWP).

Gyawali, D. & Dixit, A. (1999). Fractured Institutions and Physical Interdependence: Challenges to Local Water Management in the Tinau River Basin, Nepal. In M. Moench, E. Caspari, & A. Dixit (eds). *Rethinking the mosaic: Investigations into local water management.* Kathmandu: Nepal Water Conservation Foundation, Kathmandu and Boulder, Colorado: Institute for Social and Environmental Transition.

Gyawali, D. (2013). Reflecting on the chasm between water punditry and water politics. *Water Alternatives*, 6(2), 177–194.

Hansson, S., Hellberg, S., & Öjendal, J. (2012). Politics and development in a transboundary watershed: The case of the Lower Mekong Basin. In J. Öjendal, S. Hansson, and S. Hellberg (eds). *Politics and development in a transboundary watershed: The case of the Lower Mekong Basin* (pp. 1–18). Dordrecht, Heidelberg, London, New York: Springer.

Hardy, L., Alberto, G., & Juana, L. (2012). Evaluation of Spain's water-energy nexus. *Int. J. Water Resour. Dev.*, 28(1), 151–170.

Hassing, J. (2009). *Integrated water resources management in action: Dialogue paper.* Paris: Unesco.
Hellegers, P., Zilberman, D. (2008). Interactions between water, energy, food and environment: Evolving perspectives and policy issues. *Water Policy*, 10(S1), 1–10.

Hoff, H. (2011). Understanding the nexus. Background Paper for the Bonn 2011 Conference: The Water, Energy and Food Security Nexus. Stockholm: Stockholm Environment Institute.

Hoff, H., Iceland, C., Granit, J., Pegram, G., & Onencan, A. (2012). The Nexus in Science and Research. Presentation for the 2012 World Water Week in Stockholm. www.worldwaterweek.org/documents/WWW_PDF/2012/Wed/Holger-Hoff-nexus-nile.pdf.

Holling, C. S. (ed) (1978). *Adaptive environmental assessment and management.* New York: John Wiley and Sons.

Huitema, D., Mostert, E., Egas, W., Moellenkamp, S., Pahl-Wostl, C., & Yalcin, R. (2009). Adaptive water governance: Assessing the institutional prescriptions of adaptive (co)management from a governance perspective and defining a research agenda. *Ecology and Society*, 14(1). www.ecologyandsociety.org/vol14/iss1/art26/.

Hussey, K. & Pittock, J. (2012). The energy-water nexus: managing the links between energy and water for a sustainable future. *Ecol. Soc.*, 17(1), 31.

Loris, A. (2012). The geography of multiple scarcities: Urban development and water problems in Lima, Peru. *Geoforum*, 43(3), 612–622.

Jeffrey, P. & Gearey, M. (2006). Integrated water resources management: Lost on the road from ambition to realisation?. *Water Science & Technology*, 53(1): 1–8.

Kaika, M. (2003). The water framework directive: A new directive for a changing social, political and economic European framework. *European Planning Studies*, 11(3), 299–316.

Khan, S. & Hanjra, M. A. (2009). Footprints of water and energy inputs in food production–Global perspectives. *Food Policy*, 34(2), 130–140.

Lawford, R., Bogardi, J., Marx, S., Jain, S., Pahl Wostl, C., Knüppe, K., Ringler, C., Lansigan, F., & Meza, F. (2013). Basin perspective on the water-energy-food security nexus. *Current Opinion in Environmental Sustainability*, 5, 607–616.

Leach, M., Scoones, I., & Stirling, A. (2010). *Dynamic sustainabilities: Technology, environment, and social justice.* Abingdon, UK: Earthscan.

Leese, M. & Meisch, S. (2015). Securitising sustainability? Questioning the "water-energy and food-security nexus". *Water Alternatives*, 8(1), 695–709.

Lele, U., Klousia-Marquis, M., & Goswami, S. (2013). Good governance for food, water and energy security. *Aquatic Procedia*, 1(0), 44–63.

Malik, R. (2002). Water-energy nexus in resource-poor economies: the Indian experience. *Water Resour. Dev.*, 18(1), 47–58.

McCartney, M. & Smakhtin, V. (2010). *Water storage in an era of climate change: Addressing the challenge of increasing rainfall variability.* Colombo, Sri Lanka: International Water Management Institute.

Mehta, L., Marshall, F., Stirling, A., Shah, E., Smith, A., Thompson, J., & Movik, S. (2007). Liquid dynamics: Challenges for sustainability in water and sanitation. STEPS Working Paper 6, Brighton: STEPS Centre.

Mehta, L., Movik, S., Bolding, J. A., Derman, B., & Manzungu, E. (2016). Introduction to the special issue: flows and practices – The politics of integrated water resources management (IWRM) in Southern Africa. *Water Alternatives*, 9(3), 389–411.

Merrey, D., Drechsel, P., de Vries, F., & Sally, H. (2005). Integrating "livelihoods" into integrated water resources management: Taking the integration paradigm to its logical next step for developing countries. *Regional Environmental Change*, 5(4), 197–204.

Mitchell, B. (1990). Patterns and implications. In B. Mitchell (ed). *Integrated water management: International experiences and perspectives* (pp. 203–218). London: Belhaven Press.

Moench, M. *et al* (2003). *Fluid mosaic: Water governance in the context of variability, uncertainty and change (A synthesis paper)*. Kathmandu: Nepal Water Conservation Foundation (NWCF) and Institute for Social and Environmental Transition (ISET-Boulder).

Molle, F. (2007). Irrigation and water policies: Trends and challenges. In L. Lebel, J. Dore, R. Daniel, & Y. S. Koma (eds). *Democratizing water governance in the Mekong Region* (pp. 9–36). Chiang Mai: Mekong Press.

Molle, F. (2008). Nirvana concepts, narratives and policy models: Insights from the water sector. *Water Alternatives*, 1(1), 131–156.

Molle, F. (2009). River-basin planning and management: The social life of a concept. *Geoforum*, 40(3), 484–494.

Molle, F., Mollinga, P., & Wester, F. (2009). Hydraulic bureaucracies and the hydraulic mission:Flows of water, flows of power. *Water Alternatives*, 3(2), 328–349.

Mueller, M. (2015). The "Nexus" as a step towards a more coherent water resource management paradigm. *Water Alternatives*, 8(1), 675–694.

Mukhtarov, F. (2009). The hegemony of integrated water resources management: A study of policy translation in England, Turkey and Kazakhstan. PhD in Environmental Sciences and Policy Budapest: Department of Environmental Sciences and Policy, Central European University.

Mushtaq, S., Maraseni, T. N., Maroulis, J., & Hafeez, M. (2009). Energy and water tradeoffs in enhancing food security: A selective international assessment. *Energy Policy*, 37(9), 3635–3644.

Newell, B., Marsh, D. M., and Sharma, D. (2011). Enhancing the resilience of the Australian National Electricity Market: Taking a systems approach in Policy Development. *Ecology and Society*, 16(2), 15. www.ecologyandsociety.org/vol16/iss2/art15/.

Opperman, J. J., Royte, J., Banks, J., Day, L.R., & Apse, C. (2011). The Penobscot River, Maine, USA: A basin-scale approach to balancing power generation and ecosystem restoration. *Ecology and Society*, 16(3), 7.

Pahl-Wostl, C. (2007). Transitions towards adaptive management of water facing climate and global change. *Water Resources Management*, 21, 49–62.

Pahl-Wostl, C., Sendzimir, J., Jeffrey, P., Aerts, J., Berkamp, G., & Cross, K. (2007). Managing change toward adaptive water management through social learning. *Ecology and Society*, 12(2), 30. www.ecologyandsociety.org/vol12/iss2/art30/.

Pahl-Wostl, C., Kabat, P., & Möltgen, J. (eds) (2008). *Adaptive and Integrated Water Management: Coping with complexity and uncertainty*. Berlin: Springer.

Parés, M. (2011). River basin management planning with participation in Europe: From contested hydro-politics to governance-beyond-the-state. *European Planning Studies*, 19(3), 457–478.

Patel, R. (2009). Food sovereignty. *The Journal of Peasant Studies*, 36(3), 663–706.

Pittock, J. (2011). National climate change policies and sustainable water management: conflicts and synergies. *Ecology and Society*, 16(2), 25. www.ecologyandsociety.org/vol16/iss2/art25/.

Pittock, J., Hussey, K., & McGlennon, S. (2013) Australian climate, energy and water policies: conflicts and synergies', *Australian Geographer*, 44(1), 3–22.

Rahaman, M.M. & Varis, O. (2005) 'Integrated water resources management: evolution, prospects and future challenges. *Sustainability: Science, Practice, & Policy*, 1(1):15–21.

Rees, J. (2013). Geography and the nexus: Presidential Address and record of the Royal Geographical Society (with IBG) AGM 2013. *The Geographical Journal*, 179(3), 279–282.

Rouillard, J. J., Benson, D., & Gain, A. K. (2014). Evaluating IWRM implementation success: Are water policies in Bangladesh enhancing adaptive capacity to climate change impacts?. *International Journal of Water Resource Development*, 30(3), 515–527.

Sabatier, P. A., Weible, C., & Ficker, J. (2005). Eras of water management in the United States: Implications for collaborative watershed approaches. In P. A. Sabatier, W. Focht, M. Lubell, Z. Trachtenberg, A. Vedlitz, & M. Matlock (eds). *Swimming upstream: Collaborative approaches to watershed management* (pp. 23–52). Cambridge, MA: MIT Press.

Saravanan, V. S., McDonald, G. T., & Mollinga, P. P. (2009). Critical review of integrated water resources management: Moving beyond polarised discourse. *Natural Resources Forum*, 33(1): 76–86.

Schmidt, J. J. & Matthews, N. (2018). From state to system: Financialization and the water-energy-food-climate nexus. *Geoforum*, 91, 151–159.

Scott, C. A., Pierce, S. A., Pasqualetti, M. J., Jones, A. L., & Montz, B. E. (2011). Policy and institutional dimensions of the water-energy nexus. *Energy Policy*, 39, 6622–6630.

Scott, C. (2011). The water-energy-climate nexus: resources and policy outlook for aquifers in Mexico. *Water Resour. Res.*, 47, W00L04.

Smajgl, A., & Ward, J. (eds) (2013). *The water-food-energy nexus in the Mekong Region: Assessing development strategies considering cross-sectoral and transboundary impacts.* New York, Heidelberg, Dordrecht, and London: Springer.

Smajgl, A., Ward, J., & Pluschke, L. (2016). The water–food–energy Nexus–Realising a new paradigm. *Journal of Hydrology*, 533, 533–540.

Shah, T., Scott, C., Kishore, A., & Sharma, A. (2003). *Energy-irrigation nexus in South Asia: Improving groundwater conservation and power sector viability.* Colombo, Sri Lanka: International Water Management Institute.

Snellen, W. B., & Schrevel, A. (2004). IWRM: For sustainable use of water; 50 years of international experience with the concept of integrated water resources management; background document to the FAO/Netherlands conference on water for food an ecosystems. The Hague, 31 January–5 February 2005 (No. 1143). Alterra.

Srivastava, S., & Mehta, L. (2014). *Not another nexus? Critical thinking on the new security convergence in energy, food, climate and water.* Brighton, UK: STEPS Centre.

Stillwell, A. S., King, C. W., Webber, M. E., Duncan, I. J., & Hardberger, A. (2011). The energy-water nexus in Texas. *Ecology and Society*, 16(1), 2. www.ecologyandsociety.org/vol16/iss1/art2/.

Swatuk, L. A. (2005). Political challenges to sustainably managing intra-basin water resources in Southern Africa: Drawing lessons from cases. In *Proceedings of a workshop on development and cooperation comparative perspective: Euphrates-Tigris and Southern Africa.* Bonn: Centre for Development Research (ZEF).

Swyngedouw, E. (2011). Interrogating post-democratization: Reclaiming egalitarian political spaces. *Political Geography*, 30(7), 370–380.

Teclaff, L. A. (1967). *The river basin in history and law.* The Hague, Netherlands: Martinus Nijhoff.

UN-DESA (2011). *World economic and social survey.* New York.

Vera, J. & Zwarteveen, M. (2008). Modernity, exclusion and resistance: Water and indigenous struggles in Peru. *Development*, 51, 114–120.

Walters, C. J. & Holling, C. S. (1990). Large-scale management experiments and learning by doing. *Ecology*, 71, 2060–2068.

Wang, J., Rothausen, S., et al. (2012). China's water-energy nexus: greenhouse-gas emissions from groundwater use for agriculture. *Environ Res. Lett.*, 7(1), 014035.

WEF (2008). *Global Risk Report.* Davos: World Economic Forum.

WEF (2009). *Thirsty energy: Water and energy in the 21st century.* World Economic Forum in partnership with Cambridge Energy Research Associates.

WEF (2011). *Water security: The water-food-energy-climate nexus.* Washington, DC: Island Press.

Wichelns, D., (2017). The water-energy-food nexus: Is the increasing attention warranted, from either a research or policy perspective? *Environmental Science & Policy*, 69, 113–123.

Williams, J., Bouzarovski, S., & Swyngedouw, E. (2014). Politicising the nexus: Nexus technologies, urban circulation, and the coproduction of water-energy. ESRC: Nexus Network Think Piece Series, Paper 001.

WWAP (World Water Assessment Programme) (2012). *The United Nations World Water Development Report 4: Managing water under uncertainty and risk*. Paris: UNESCO.

Xie, M. (2006). *Integrated water resources management (IWRM): Introduction to principles and practices*. New York: World Bank Institute.

Zwarteveen, M. Z. & Boelens, R. (2014). Defining, researching and struggling for water justice: Some conceptual building blocks for research and action. *Water International*, 39(2), 143–158.

4

THE KNOWLEDGE NEXUS AND TRANSDISCIPLINARITY

4.1. Why nexused interdisciplinarity?

Recent decades have witnessed the emergence of various movements concentrating on interconnections among nature, society, and technology – and the disciplines that deal with them – whether under the broad title of environmental science, or more specific intellectual traditions like transition theory (Geels 2002; Loorbach, Frantzeskaki, & Avelino 2017; Schwanen 2018). This compulsion is even more prevalent among resource managers who have to balance competing claims for contradictory ends. Water managers, for instance, have long wrestled with the uncomfortable fact that water is not so much a subject of study or praxis but more realistically a focal point where just about every subject taught in a university's different departments intersect: from atmospheric physics to hydrogeology, from civil engineering to economics, law, sociology, politics, ethics, and even literature (Gyawali 2010). How to solve a water problem facing a business, community, or municipality without running into opposition from competing claimants or disciplines has been a vexing and perennial problem. Efforts to address this difficulty is what led to the emergence of the movement for Integrated Water Resources Management (IWRM), itself a successor to the earlier approach which was the hydrology-inspired river basin management of the 1960s (Chapter 3). Allan (2003) has argued that IWRM too is failing since its votaries are not recognising that it is broader than water and environment, that both integration and management are political processes which requires greater disciplinary ecumenism than practiced currently.

The nexus represents the latest in this evolving series of paradigms, described as a multi-dimensional means of scientific enquiry which seeks to describe the complex and non-linear interactions between water, energy, food, with the climate, and further understand wider implications for society. The shift to a nexus approach also parallels the shift from the much narrower Millennium Development Goals

(MDGs) to the current Sustainable Development Goals (SDGs) since 2015, which are seen as more comprehensive and addressing concerns such as energy and equity missed by the MDGs. However, this encompassing of broader concerns brings with it significant conceptual, methodological, and practical issues: how comprehensive can one be without losing analytical rigour; what tools need to be used; and what are the practical consequences? In addressing these conundrums, the nexus is emerging as the epicentre, or meeting point of a series of (often complex) components, which come together to represent something that is more than the sum of its parts. As a result, interdisciplinary debates on the nexus focus on: (i) what it is that is "connected"; (ii) the exact nature of those connections; and (iii) boundary issues, i.e. if everything is linked in some way, then when and where do we draw the line (Howarth & Monasterolo 2017)?

Thus far, however, specific methods to address complex resource interactions with development challenges remain limited. There are of course specific data issues which make nexus research more difficult, in particular the lack of data and limited data interoperability (on lack of data, see Houghton-Carr, Fry, & Wallingford 2006; on limited data interoperability, see Mohtar & Lawford 2016 and Eftelioglu et al 2017). Lack of data, limited data interoperability, and data incompatibility are a few of the many data challenges hindering meaningful integration of relevant nexus data. Integrative and interdisciplinary frameworks and models are needed to create compatible datasets, which will then be able to support decision-making, for example through interactive platforms and maps. The current available methodologies present many issues. A systematic review of 245 journal articles and book chapters reveals that (a) use of specific and reproducible methods for nexus assessment is uncommon (less than one-third); (b) nexus methods frequently fall short of capturing interactions among water, energy, and food – the very linkages they conceptually purport to address; (c) assessments strongly favour quantitative approaches (nearly three-quarters); (d) use of social science methods is limited (approximately one-quarter); and (e) many nexus methods are confined to disciplinary silos – only about one-quarter combine methods from diverse disciplines and less than one-fifth utilise both quantitative and qualitative approaches (Albrecht, Crootof, & Scott 2018). Most methodological approaches on the nexus are biased towards quantitative methods (more than 70 per cent of studies used primarily quantitative approaches), mostly life-cycle assessment, input-output analysis, trade-off analysis, foot printing, or integrated models with scenario analysis (Albrecht, Crootof, & Scott 2018). Therefore, nexus-specific methods that better represent cross-sectoral social, environmental, and technical challenges are needed, as discussed in this chapter.

Since 2011, one can witness a proliferation of research articles on the need for interdisciplinarity and transdisciplinarity when approaching the nexus, reflecting a call for greater integration of research efforts and policy prescriptions across disciplines (Endo et al 2015a; Stirling 2015; Wichelns 2017). Many researchers and policy analysts would consider that interactions across disciplines and sectors generally are helpful and informative. However, this is not a one-way debate, many also suggest there are many instances in which problem solving even within a complex system requires sharp focus within a discipline and narrow, high-level expertise that comes with it (Jones 2009; Kanakia 2007). In such instances, problems are best solved, and certain problems of efficiency or equity are best handled within policies that are narrow and with less than complete discussions across sectors.

The concept of the nexus is not just a policy tool and function but also an approach that requires a transformation in the way disciplines and sectors are working. The knowledge nexus will require re-thinking interdisciplinarity. Different organising styles – governments, markets, or movements – have different institutional filters that allow some data in as information but filter out others as "noise" (Gyawali et al 2006). Understanding this dynamic is critical to the nexus approach where interdisciplinarity has to be re-thought in terms of whether only one hegemonic discipline "feeds" the problem or is the feeding more plural. What is needed, is not only "joined up thinking", but profoundly transformative change in infrastructures, organisations, behaviours, markets, governance practices, and even cultures more widely. These are the challenges linked to what we have termed the knowledge nexus.

The knowledge nexus is about a transdisciplinary approach, aimed at opening up and broadening out analysis, and linking theory and method to practical solutions that address today's global sustainability challenges. We will go beyond multi-disciplinarity (combining disciplines) and interdisciplinarity (joining disciplines), to a process of joint learning and co-production. This requires broad-based research (across disciplines and methods), and co-creation of solutions (across sectors and including citizens). Through enabling (rather than suppressing) scepticism and criticism, policies become more robust, responsible, and accountable. Messy, bottom-up transdisciplinarity can yield unexpected insights and possibilities, through exposure to other kinds of tacit, non-specialist, or general knowledge – held by local communities, businesses, social movements, or many different kinds of practitioners (Stirling 2015). Through facilitating more radical interactions between different styles of knowledge, potentially transformative solutions can be fostered.

4.2. Interdisciplinarity and transdisciplinarity

Foran (2015) has explained two broad disciplinary traditions within nexus research. The first might be summarised as the systems complexity of the nexus, more specifically the systematic connections between domains (e.g. food production). Disciplines such as economics, hydrology, engineering life cycle analysis, scenario analysis, and systems analysis have been used to describe such connections (Bazilian et al 2011; Hoff 2011; Howells et al 2013; Hussey & Pittock 2012; Newell, Marsh, & Sharma 2011). Findings are conveyed in terms of efficiency, productivity, trade-offs, synergies, and co-benefits.

The second tradition is around critical social science of the nexus, with a focus on power relations and the historical, cultural, and socio-political dimensions of these relationships. This topic raises questions such as: "[h]ow has the resource nexus in a particular place emerged, historically? Which social groups are enriched

TABLE 4.1 Two approaches towards the resource nexus

Characteristic properties	Complex systems thinking	Critical social sciences
Focus	Cross-level, cross-domain impacts of particular actions	Historical determinants of vulnerability, insecurity, or poverty in specific places Winners and losers from particular actions
Key processes	Absolute limits (biophysical, social) Interactions between reinforcing (positive) and balancing (negative) feedback Cross domain interactions Unintended consequences Learning	Capitalist accumulation Market imperative Dispossession Institutions Discursive power Differences and stratification (e.g. gender, caste, class)
Common sequence of analysis	Macro - > Meso - > Macro	Micro - > Meso - > Macro
Specific techniques	Quantitative modelling Scenario analysis	Historical analysis Critical discourse analysis In-depth actor interviews Ethnography

Source: Foran 2015: 658

(impoverished) by a particular resource nexus? Who gains or losses from attempts to intervene in the nexus?" (Allouche, Middleton, & Gyawali 2015; Barney 2012; Foran & Manorom 2009; Friend, Arthur, & Keskinen 2009; Molle et al 2009).

A holistic understanding of complex phenomena such as the resource nexus requires some kind of interdisciplinary inquiry. Because the two approaches differ in focus, theoretical processes, typical sequence of analysis, and techniques, combining them is analytically intensive, and presents challenges of epistemology (Foran 2015). This is true of practitioners as well as theoreticians. Gyawali (1989), summarising the experiences of a water engineer thrust into a policy environment, distinguishes a multidisciplinary approach from an interdisciplinary one. The former consists of interdepartmental commissions and task forces where experts from different disciplines are brought together and contribute to the analysis of that slice of the overall problem at hand which are amenable to the application of their disciplinary tools and that ignore other aspects that lie in the penumbra of inter-linkages. In this approach, the final synthesising of the various disciplinary solutions is left to some harried politician-minister with little training for it.

In contrast, an interdisciplinary inquiry strives to use the concerns of other disciplines to re-structure the arguments of one's own discipline. A common example from the water sector is when engineers planning a dam have to re-examine their technically optimal design with considerations of economics (can it be paid for?) or sociology (will those to be relocated agree to be displaced or not or at what price, which may require changing the dam height?) or law (how many court cases will we have to face from angry activists?). Breaking a whole into disciplinary components is essential for analysis, which is the act of *describing* different facets of a complex problem: when one wishes to do something about such a vexing problem, i.e. prescribe a policy for action in *solving* the problem, one needs to synthesise the various knotty aspects of it. That summing up for action is the essence of interdisciplinarity, which at its core is also what political decision-making is all about. Such enlightened formula for the use of power, i.e. policy, is an area of academic enquiry that is still in its infancy since established universities and departments tend to reward disciplinary contributions rather than interdisciplinary ones.

However, most researchers argue that interdisciplinarity is not enough, and what is required for the nexus is transdisciplinarity. There are numerous definitions of transdisciplinarity and understandings of how this differs from multi and interdisciplinarity. Barry, Born, and Weskalnys (2008: 28) view these as a spectrum from multidisciplinarity – cooperation of disciplines whose framings remain largely intact – to transdisciplinarity. The latter term captures a type of reflexive and integrative knowledge production that is oriented at application and addressing societal and environmental problems and involves non-academic stakeholders as active participants (Klenk & Meehan 2015; Osborne 2015). Transdisciplinary methods aim for broad participation and to incorporate knowledge from various sources, such as academic research, on-the-ground practitioner experience, and local knowledge (Mauser et al 2013). By participating in the research process, stakeholders help guide the research questions, and study design. Transdisciplinary approaches are used to identify inter-sectoral relationships, achieve more holistic assessments, and improve integration of policy among sectors (Endo et al 2015b). Transdisciplinarity relates scientific and societal problems and "produces new knowledge by integrating different scientific and extra-scientific insights; its aim is to contribute to both societal and scientific progress" (Jahn, Bergmann, & Keil 2012: 8).

Harris and Lyon (2014) defined the requirements of a transdisciplinary approach for nexus analysis according to a literature review on transdisciplinary research. In particular, they identified the associated theoretical (framing problems), methodological (different conceptions of proof), and practical challenges (communication, collaboration, and trust across groups of actors belonging to different disciplines) for nexus analysis. For academia, they found that a key challenge relates to the need to embrace multidimensional knowledge, and to adapt the method of interaction to account for transdisciplinary team members (e.g. defining a new language, negotiate, accept the different logics and values, redefine the power balances among disciplines and among scientists and lay practitioners).

Thoren (2015) discusses interdisciplinarity in the context of sustainability science, an emerging field aimed at challenges like global warming, which hopes to lay the foundations for a form of science that serves society by solving "real world problems" leading towards planetary sustainability. Its strategy is to draw on the resources of a wide range of disciplines, both of natural and social sciences; but the challenge before that task is that of integration and joint problem solving. According to Thoren (2015), interdisciplinarity is often used as an umbrella term capturing many types of relationships between disciplines; but they can be reduced to a trichotomy of multidisciplinarity, interdisciplinarity, and transdisciplinarity. Multidisciplinarity is understood here as merely the juxtaposition of knowledge claims from different disciplines that is additive and not integrative. By contrast, interdisciplinarity is integrative and, drawing on Klein (2010), Thoren distils the process of integration to: borrowing of tools and methods; solving problems without necessarily achieving conceptual unification of knowledge; increased consistency of subjects and methods nudging disciplines to a state of merger; and finally the emergence of hybrid disciplines at the fringes or interface of already existing ones such as social psychology. Finally, for Thoren, transdisciplinarity, a term that emerged at an OECD conference in 1970, is the final and deepest stage of interdisciplinary collaboration with complete coordination, a kind of overarching synthesis.

This kind of broad transdisciplinary synthesis is guided by two features. The first is integration of science with society, including local knowledge and value systems, to solve real world problems where uncertainties are high and the underlying values are being challenged, whilst solutions are urgent. The second is problem orientedness, where problems to be solved originate in the real world and not in the highly idealised theoretical frameworks or sterile laboratories. It is here that Thoren (2015) brings out the idea of "problem feeding" that distinguishes transdisciplinarity from multidisciplinarity and interdisciplinarity and which involves problem sharing and problem transfer between disciplines. And it is with this idea of problem feeding that transdisciplinarity becomes germane to the nexus approach. As a concept, the nexus is supported by a rapidly growing evidence base and a community of practitioners and policy makers, providing a powerful but largely disconnected knowledge base to understand the relationships and trade-offs between the different sectors and disciplines characterising the nexus (Azapagic 2015; Harris & Lyon 2014; Kurian & Ardakanian 2014; Stirling 2015).

4.3. Further transdisciplinary work required

There is a normative danger to recognise when engaging with interdisciplinarity and transdisciplinarity. Thinking interconnections across domains or systems is difficult, the argument goes, because of the specialisation and fragmentation of science. Howarth and Monasterolo (2016) for instance argue that a nexus approach enables the capitalisation of knowledge and the sharing of skills and expertise to build innovative solutions to complex interlinked nexus challenges. Much of the literature promoting the water–food–energy nexus is somewhat dismissive of research or policy analysis conducted in "silos", in which scientists or public officials pursue narrowly focused inquiries, without sufficiently interacting with specialists across technical disciplines (Azapagic 2015; De Laurentiis, Hunt, & Rogers 2016; Finley & Seiber 2014; Leck et al 2015; Sharmina et al 2016). None of the authors expressing this perspective provides evidence of any harm that has arisen as a result of such analysis. Yet, the implication seems to be that a nexus approach is superior to conducting research and policy analysis within scientific disciplines

(Wichelns 2017). Interdisciplinarity can be pursued for different reasons (Barry, Born, & Weskalnys 2008) but with transdisciplinarity the most common reason remains integration. This is underpinned by a presumption of superiority: knowledge production can be improved and made more effective and impactful in addressing societal problems if the inevitable partial and telescopic character of disciplinary perspectives and practices is overcome through some kind of fusion.

4.4. Interdisciplinarity and transdisciplinarity as transformative?

The potential for interdisciplinarity and transdisciplinarity to deliver innovation in its widest sense is large. Innovative transdisciplinary approaches are being increasingly used to address important societal challenges (Bammer 2013) and facilitate and navigate the interrelationships and trade-offs between energy, food, and water within the nexus in parallel to the varying and often conflicting needs of actors involved (Polk 2014; Zhang & Vesselinov 2016). However, there is always the danger that new research "performed" under this nexus transdisciplinary movement is not innovative at all, and just repeats and reaffirms old established knowledge.

One such example is the case of Hindu Kush Himalayan ecosystem services in South Asia, which demonstrated that in order to sustain resilience of resources and water, food, and energy security in the region, cross-sectoral integration was needed, along with regional integration between upstream and downstream players, critical for ensuring water, food, and energy security (Rasul 2014). Another example, is the context of sustainable consumption of food, water, and energy, a practices approach would explore the social organisation of cooking, which, as an activity, consumes food, water, and energy, and can complement more traditional approaches in sociology. Similarly exploring the full impacts of a complete food chain through life cycle thinking (Azapagic 2015) could increase understanding of the diverse mechanisms that could be used to reduce the impact of this sector on exacerbating nexus shocks such as climate change (Jeswani, Burkinshaw, & Azapagic 2015). The point is not to be critical about these research projects, as these conclusions are perfectly sensible. However, they reveal the danger that a reframing of the problem through the knowledge nexus and transdisciplinarity may not be as transformative without a higher level of integration establishing a common system of axioms for a set of disciplines.

4.5. Policy or politics: The road to more policy toolkits?

The second danger of this call for a transdisciplinary approach is to render it technical. Policy rather than politics becomes the focus, and it becomes a research governance question from which implications for policy making are derived. It becomes narrowly framed as a way to consider complexities around the variety and forms of data used to inform nexus-related decision making (Gilbert & Bullock 2014). A number of studies have focused on these policy question. Howarth and Monasterolo (2016: 17), for example, have developed a transdisciplinary approach to knowledge development through co-production on water–food–energy nexus decision-making in the UK. Co-production as a methodology provides a space to facilitate knowledge exchange and sharing of insights from a range of perspectives and expertise, acknowledging that all those who contribute to the process have something to offer. It enables an inclusive, self-reflective approach whilst embracing the challenges that the process faces – and acknowledging the opportunities this provides (Howarth & Monasterolo 2017). In their knowledge co-production approach, expertise was drawn upon from across disciplines and fields as represented by the diversity of individuals invited to take part in the workshops. The multidimensional methodological framework was designed to account for feedback loops and cascading effects, and sought to inform decision-making processes to build societal resilience to nexus shocks going beyond the sectorality of current research practice (Howarth & Monasterolo 2016).

Howarth and Monasterolo's (2016) analysis of workshop discussions identified four dominant themes that emerged as barriers to decision-making in the context of nexus shocks: communication and collaboration, decision making processes, social and cultural dimensions, and the nature of responses to nexus shocks. Communication and collaboration are seen as vital to ensure the most appropriate and robust evidence informs decision makers at all levels within the context of a nexus shock. For example, collaboration between actors across sectors can lead to clashes in languages and lexicons as well as skillsets and expertise further exacerbating barriers that may emerge in the communication process (Howarth & Monasterolo 2016).

One important concern that was identified by the workshop participants was the potential tension between probabilities and levels of uncertainty and clear advice for decision makers. In terms of decision-making processes, workshop participants highlighted the lack of clarity over who owns the problem or the decision. Conflicting timescales between research and policy combined with the social dimensions of decision-making and the need for researchers to achieve consensus before they can contribute to decision making can exacerbate responses to shocks and cause existing decision-making processes to become redundant.

A second issue shared by workshop participants was the lack of learning systems in place, "to capture these lessons during and after the shock, how this could inform thinking in future shocks and how these lessons learnt could then be transferred and applied to other sectors and scales" (Howarth & Monasterolo 2016: 57). The third theme highlighted was the different cultures, behaviours, priorities, and processes by different stakeholders across different sectors. Finally, in terms of response to shock, the production of scientific evidence used to inform decision-making is imprecise, fraught with uncertainties, and constantly evolving. According to Howarth and Monasterolo (2016), the need to move from the current reactive to a proactive decision-making process emerges strongly, with a necessity to embrace a foreseeing attitude to future nexus shocks and understand the importance of local action for global impacts.

These four themes and their findings are interesting but reveal a particular framing of how transdiciplinarity is being conducted. It is about closing down policy options and reaching a consensus to support decision-making. The danger is that nexus transdisciplinary research becomes stuck in producing "tools", "techniques", and "frameworks" for policy making, as illustrated in the logic above and Table 4.2. Here, academics and scientists offer putatively neutral, a-political information that will allow others – politicians, policy-makers, businesses – to make decisions that help to increase the efficiency of resource allocation. In the process, they may miss transforming the previously unsustainable politics into a new sustainable one (Scoones, Leach, & Newell 2015).

While much nexus research aspires to be post-normal and transdisciplinary, a lot of work therefore remains within a conventional model for stakeholder interactions that seeks to separate facts from values. It also reveals the lack of interdisciplinary

TABLE 4.2 Perceived opportunities to increase resilience to nexus shocks

Contextual factors that help mitigate nexus shocks and include (i) the importance of clarifying what we consider as a cost, differentiating between computable, perceived, and opportunity costs, and between costs that could be afforded (financial) or not (human lives) in case of shocks; (ii) the emergence of a strong internal leadership; (iii) the increasing collaboration between stakeholders and across sectors as well as transparency and information sharing; (iv) the improving communication of evidence and impacts of shocks targeting the language to the specific audience.

Strategic thinking that builds on the understanding of the big picture of nexus shocks' complexity and consists of (i) having a context-specific plan B to react quickly to nexus shocks prioritising interventions based on lessons learned from previous experiences; (ii) clear division or roles to allow clear identification of interlocutors and match policy response to shocks; (iii) decentralisation of decision-making and shared responsibility to increase stakeholders' engagement and ownership of responses to shocks.

Collaboration and communication characterised by the importance of establishing knowledge-transfer partnerships to design and implement a robust and efficient response to shocks by better understanding the longer-term risks associated with nexus shocks and building nexus narratives and framing responses with a focus on opportunities and business solutions. Moreover, the creation of a common stakeholders' language and narrative around nexus shocks is important to coordinate responses.

Anticipating social responses, by blending insights from the multiple sectors involved in the response to nexus shocks thus complementing knowledge and providing a framework which considers the big picture, to better deal with the complexity of nexus shocks. In this regard, it is fundamental to increase the accountability of the decision-making process combining evidence and data from decision makers and narrowing the gap between short-term policy objectives and long-term frameworks of measures to manage nexus shocks.

Processes to shape the right governance structure to respond to nexus shocks with the following desirable characteristics: (i) resilience and efficiency to enable flexible planning and procedures, (ii) complementary and flexible mechanisms and institutions able to operate swiftly when needed, and (iii) innovation to decentralise decision-making to better manage tailored, case-by-case solutions to cope with nexus shocks.

The relevance of **proper timescales in decision making** emerged as a transversal opportunity in all the five themes.

Source: Howarth & Monasterolo 2017: 107

FIGURE 4.1 Word cloud
Source: Based on Howarth & Monasterolo 2016

research with critical social sciences as discussed earlier. What is happening is that transdisciplinary policy and consensual-type research in relation to the nexus is become disciplined and self-regulated, where members share similar epistemic cultures and research fields with specific concerns, methods, vocabularies, and institutions. In other words, a complex landscape of power relations and forces within academia affects nexus transdisciplinary research.

The following discourse analysis through a word cloud software of Howarth and Monasterolo's (2016) article shows the emerging consensual dominant vocabulary and lexicon.

Actors, responses, and decisions become the main underpinning beyond nexus transdisciplinary research. Current policy making debates and political imperatives around the nexus are challenging transdisciplinary research through pressures to claim a definitive basis for predictive explanation of causal dynamics at a sufficient level of confidence and precision to justify large business strategies, infrastructure investments, and long-term policy commitments.

As suggested by Schwanen (2018), there is perhaps a need to slow down trans and interdisciplinary reasoning and practices. Particular concepts, ideas, logics, and methods should not be plugged as ready-mades; they will have to be adapted and hybridised to a greater or lesser degree and may even have to be dropped altogether. In line with Stengers (2011), transdisciplinary nexus research should aim to understand the world not as a messy realm of competing value systems from which research should abstract to arrive at transcendental and disinterested truths, but as an inevitable condition they have to appreciate and learn from. Research practices should offer spaces for friction by allowing competent colleagues and non-academics to object and induce other modes of thinking. Interdisciplinarity and transdisciplinarity are about experiencing and dealing with contact zones as "social spaces where cultures meet, clash, and grapple with each other, often in context of highly asymmetrical relations of power" (Pratt 1991: 34). Yet, if asymmetrical exchange of concepts, ideas, and methods across disciplines, epistemic communities, and research fields and fragmenting pluralism are seen as undesirable, then a slowing down of reasoning will be needed (Schwanen 2018).

One method of conceptualising the inherently plural nature of social interactions is through the neo-Durkheimian Theory of Plural Rationalities (or as more popularly known as Cultural Theory, see Beck et al 2018; Thompson 2008; Verweij and Thompson 2006). This integrative social science argues that, with just two discriminators showing whether competition is fettered or unfettered and whether transactions are symmetrical or asymmetrical, there emerge four styles of organising: the first three being active bureaucratic hierarchism of procedural rationality, market individualism of substantive rationality, activist egalitarianism of critical rationality and the fourth being voter and consumer fatalism based on passive coping rationality. Each of them upholds a different view of nature (nature robust within limits, nature robust, nature fragile, and nature capricious) as well as different approach to risk (risk managing, risk taking, risk amplifying, and risk coping respectively). Within this framing of social interactions (it is the first three active social styles of organising that strategise and cognise, strive to disorganise the others, and seek to bring into its fold of sanctioned behaviour the passive voters and consumers), problem definition itself becomes plural and hence proposed solutions even more varied. Thus, transdisciplinarity would mean that all the voices be not only heard at the policy table – which should not be hegemonied by the state agencies or that of the market – but also responded to. This is what in essence is the purpose and meaning of "problem feeding" discussed above. It is not of much use if different voices are heard but the others doing the hearing do not internalise those concerns and do not respond to them. This listening to other organising styles, re-examining their concerns, and responding with revised options is then what transdisciplinarity would mean.

Cultural Theory would also posit that such a "constructive engagement" between different styles of organising is what would ultimately force a nexused understanding among them. Indeed, bureaucratic hierarchism would veer towards procedural solutions of laws and regulations, market individualism would prefer to listen to neoliberal economics of efficiency and profit, while activist egalitarianism would opt for critical social sciences that bring equity and justice to the forefront. All three would have their strengths and weaknesses, all three would have a grasp of some aspect of the complex socio-environmental reality but none of them would be wholly right either. If the engagement among them in the policy terrain in not one of hegemony but democratic listening and responding, then a better nexused common and integrated understanding would have been achieved.

So, it is important to recall that nexus-related challenges can be more about enabling empowering hopes for distributed social progress, than urgent, top-down assertions of catastrophic technical fears. Key progressive responses to global challenges in achieving equitable and sustainable provision of water, food, and energy are not about "sound scientific" research informing "evidence-based policy" to enable "pro innovation" strategies that roll out global programmes for "scaling up" the diffusion of particular "technological solutions". These familiar kinds of high level policy buzzword do not just present too simple a picture. They also inflect it in highly partisan political ways, of a kind that are arguably more aligned with sustaining existing structures of privilege than achieving real material progress in addressing nexus-related challenges. In short, they treat nexus-related progress as a matter primarily of elite experts successfully engaging with elite policy makers (Stirling 2015).

How to go beyond narrow risk-based methods of "sound scientific", "evidence-based policy" to more fully address conflicting values, uncertainty, ambiguity, and outright ignorance? The nexus is characterised by high levels of interconnectivity and uncertainty (Howarth & Monasterolo 2016). Nexus-related interactions involve many different kinds of processes and relations, typically changing in highly dynamic ways. This means that consequences of different conditions and interventions are typically nonlinear – not only unpredictable but often profoundly surprising in ways that defy conventional statistical forecasting, optimising calculations, or aggregating models. The water, food, energy nexus is a dauntingly deep and pervasive constellation of interacting global, natural, social, and technological systems. That any given methodology might confidently yield even a generally robust appreciative understanding of key drivers and patterns in any given context, would be misleading (Stirling 2015). It is for this reason that Klenk and Meehan (2015) advocate a mode of transdisciplinary research on environmental issues that values difference and recognises that the construction of a knowledge nexus is unequal and power-laden.

So, what are the practical implications for concrete nexus-focused methods and methodologies of this more explicit and realistic recognition of the roles played by power asymmetries in scientific research and knowledge production more generally? Before considering more general cross-cutting challenges in nexus-related methods and capabilities, the main point here is one of radical diversity. It is not just robust decision-making and democratic accountability that provide imperatives for seeking alternative practical methods for addressing nexus-related challenges. Scientific rigour also demands more open-ended forms of uncertainty heuristics, interval analysis, sensitivity testing, and scenario assessment – each requiring attention to the differing conditions that may frame the question at hand (Stirling 2015).

One way of summarising the methodological implications of uncertainty, ambiguity, and ignorance, is that they establish the need to radically "broaden out" and "open up" the range and kinds of methods used to produce knowledges about water, food, and energy nexus linkages and interventions. In other words, there is a premium on those particular tools, techniques, and frameworks, that are capable of taking into account a wider range of interacting factors in nexus-related challenges, scrutinise a more complete array of possible policy interventions, and engage with a greater diversity of ways of understanding these. In order to provide a basis for decision-making that is as robust as possible, this evidence and analysis should be communicated with policy debates and wider political arenas in ways that are as systematic, clear and transparent as possible about contestable implications (Stirling 2015).

So crucial kinds of capacity-building for effective nexus-focused methodologies lie in nurturing capabilities that directly resist and counter any uneven balance of power. These include: egalitarianism, humility, pluralism, and reflexivity on the part of all communities involved in nexus-related research and appraisal (Stirling 2015).

Egalitarianism means that practical implementation of nexus-related methods does not simply assume and apply the particular questions, framing assumptions, priority values, or boundary conditions asserted by the loudest or highest status "users". Here (whether methods are interactive or analytic, quantitative or qualitative), training and skills for design and conduct of nexus-focused research and appraisal require capabilities to interrogate and more fairly counterbalance such bias and privilege.

Humility requires the building of capabilities among those institutions and disciplines benefiting from established structures of privilege in nexus-related appraisal, enabling them to be more deliberate in creating spaces for others – not denying contrasting understandings as "irrational", "ignorant", or "jargonistic". This means a readiness to be led where appropriate by agendas or questions set outside a particular home discipline or beyond academic disciplines entirely.

Pluralism requires an ethic of tolerance for interests, values, or knowledges that are not only different, but directly contending with those of a particular individual, organisation, or discipline. It means a capability to express and respond to scepticism, without interpreting this as existential denial. By encouraging (rather than suppressing) critical discourse, this helps foster more robust knowledge.

Finally, reflexivity is a quality whose very recognition requires all the above capabilities. It is the further more demanding capability to acknowledge how nexus-related challenges can look fundamentally different depending on the perspective from which they are viewed. This arises especially (though not exclusively) from critical social science analysis, since this (by definition) involves interrogation and distancing from conventional interests and assumptions. But reflexivity is more a relational capability among interacting groups of perspectives, than a transcendent virtue located in any particular framework or individual.

References

Albrecht, T. R., Crootof, A., & Scott, C. A. (2018). The water-energy-food-nexus: A systematic review of methods for nexus assessment. *Environ. Res. Lett*, 13., in press.

Allan, T. (2003). IWRM/IWRAM: A new sanctioned discourse? Occasional Paper 50, SOAS Water Issues Study Group. London: School of Oriental and African Studies/King's College London.

Allouche, J., Middleton, C., & Gyawali, D. (2015). Technical veil, hidden politics: Interrogating the power linkages behind the nexus. *Water Alternatives*, 8(1), 610–626. Downloadable at: www.water-alternatives.org/index.php/alldoc/articles/vol8/v8issue1/277-a 8-1-1/file.

Azapagic, A. (2015). Special issue: Sustainability issues in the food-energy-water nexus. *Sustain. Prod. Consum.*, 2, 1–2.

Bammer, G. (2013). *Disciplining interdisciplinarity: Integration and implementation sciences for researching complex real world problems*. Canberra, Australia: ANU Press.

Barney, K. (2012). Land, livelihoods and remittances: A political ecology of youth out-migration across the Lao-Thai Mekong Border. *Critical Asian Studies*, 44(1), 57–83.

Barry, A., Born, G., & Weskalnys, G. (2008). Logics of interdisciplinarity. *Theory, Culture and Society*, 37, 20–49.

Bazilian, M., Rogner, H., Howells, M., Hermann, S., Arent, D., Gielen, D., & Yumkella, K. K. (2011). Considering the energy, water and food nexus: towards an integrated modelling approach. *Energy Policy*, 39(12), 7896–7906.

Beck, M. B., Thompson, M., Gyawali, D., Langan, S., & Linnerooth-Bayer, J. (2018). Viewpoint – Pouring money down the drain: Can we break the habit by reconceiving wastes as resources?. *Water Alternatives*, 11(2), 260–283.

De Laurentiis, V., Hunt, D.V., & Rogers, C.D. (2016). Overcoming food security challenges within an energy/water/food nexus (EWFN) approach. *Sustainability*, 8, 95.

Eftelioglu, E., Jiang, Z., Tang, X., & Shekhar, S. (2017). The nexus of food, energy, and water resources: Visions and challenges in spatial computing. In *Advances in geocomputation* (pp. 5–20). Switzerland: Springer.

Endo, A., Tsuritab, I., Burnett, K., & Oencio, P. M. (2015a). A review of the current state of research on the water, energy, and food nexus. *J. Hydrol. Reg. Stud.*, 11, 20–30.

Endo, A., Burnett, K., Orencio, P. M., Kumazawa, T., Wada, C. A., Ishii, A., Tsurita, I., & Taniguchi, M. (2015b). Methods of the water-energy-food nexus. *Water*, 7, 5806–5830.

Finley, J. W. & Seiber, J. N. (2014). The nexus of food, energy, and water. *J. Agric. Food Chem.*, 62, 6255–6262.

Foran, T. (2015). Node and regime: Interdisciplinary analysis of water-energy-food nexus in the Mekong region. *Water Alternatives*, 8(1), 655–674.

Foran, T. & Manorom, K. (2009). Pak Mun Dam: Perpetually contested? In F. Molle, T. Foran, & M. Käkönen (eds). *Contested waterscapes in the Mekong Region: Hydropower, livelihoods and governance*. London: Earthscan.

Friend, R., Arthur, R., & Keskinen, M. (2009). Songs of the doomed: The continuing neglect of capture fisheries in hydropower development in the Mekong. In F. Molle, T. Foran, & M. Käkönen (eds). *Contested waterscapes in the Mekong region: Hydropower, livelihoods, and governance*. London: Earthscan.

Geels, F. W. (2002). Technological transitions as evolutionary reconfiguration processes: A multi-level perspective and a case-study. *Research Policy*, 31, 1257–1274.

Gilbert, N. & Bullock, S. (2014). Complexity at the social science interface. *Complexity*, 19 (6), 1–4.

Gyawali, D. (2010). What is Special about Water? In J. Dore, J. Robinson, & M. Smith (eds) *Negotiate: Reaching agreements over water*. Gland, Switzerland: IUCN.

Gyawali, D., Allan, J. A., Antunes, P., Dudeen, A., Laureano, P., Fernández, C. L., Scheel Monteiro, P. M., Nguyen, H. K., Novácek, P, & Pahl-Wostl, C. (2006). *EU-INCO water research from FP4 to FP6 (1994–2006): A critical review*. Luxembourg: Office for Official Publications of the European Communities.

Gyawali, D. (1989). *Water in Nepal*. Occasional Paper No. 8. Hawaii: East-West Center Environment and Policy Institute. Reprinted in Gyawali, D. (2003). *River, technology and society: Learning the lessons of water management in Nepal*, London: Zed Books.

Harris, F. & Lyon, F. (2014). *Transdisciplinary environmental research: A review of approaches to knowledge co-production*. Brighton, UK: Thinkpiece Series.

Hoff, H. (2011). Understanding the nexus. Background Paper for the Bonn 2011 Conference: The Water, Energy and Food Security Nexus. Stockholm: Stockholm Environment Institute.

Houghton-Carr, H., Fry, M., & Wallingford, U. K. (2006). The decline of hydrological data collection for development of integrated water resource management tools in Southern Africa. *IAHS publication*, 308, 51.

Howarth, C. & Monasterolo, I. (2016). Understanding barriers to decision making in the UK energy-food-water nexus: The added value of interdisciplinary approaches. *Environmental Science & Policy*, 61, 53–60.

Howarth, C. & Monasterolo, I. (2017). Opportunities for knowledge co-production across the energy-food-water nexus: Making interdisciplinary approaches work for better climate decision making. *Environmental Science & Policy*, 75, 103–110.

Howells, M., Hermann, S., Welsch, M., Bazilian, M., Segerstrom, R., Alfstad, T., Gielen, D., Rogner, H., Fischer, G., Velthuizen, H. van., Wiberg, D., Young, C., Roehrl, R. A., Mueller, A., Steduto, P., & Ramma, I. (2013). Integrated analysis of climate change, land-use, energy and water strategies. *Nature Clim. Change*, 3(7), 621–626.

Hussey, K. & Pittock, J. (2012). The energy–water nexus: Managing the links between energy and water for a sustainable future. *Ecology and Society*, 17(1), 31.

Kurian, M. & Ardakanian, R. (eds). (2014). *Governing the nexus: Water, soil and waste resources considering global change*. Switzerland: Springer International Publishing.

Leck, H., Conway, D., Bradshaw, M., & Rees, J. (2015). Tracing the water–energy–food nexus: description, theory and practice. *Geography Compass*, 9(8), 445–460.

Loorbach, D., Frantzeskaki, N., & Avelino, F. (2017). Sustainability transitions research: Transforming science and practice for societal change. *Annual Review of Energy and Resources*, 42, 4. 1–4. 28.

Jahn, T., Bergmann, T., & Keil, F. (2012). Transdisciplinarity: Between mainstream and marginalization. *Ecological Economics*, 79, 1–10.

Jeswani, H. K., Burkinshaw, R., & Azapagic, A. (2015). Environmental sustainability issues in the food–energy–water nexus: Breakfast cereals and snacks. *Sustainable Production and Consumption*, 2, 17–28.

Jones, C. (2009) Interdisciplinary approach: Advantages, disadvantages, and the future benefits of interdisciplinary studies. *ESSAI*, 7, Article 26. Available at: http://dc.cod.edu/essa i/vol7/iss1/26.

Kanakia, R. (2007). Talks touts benefits of interdisciplinary approach, as well as some of its pitfalls. Stanford Report. http://news-service.stanford.edu/news/2007/february7/barr-020707.html.

Klein, J. T. (2010). A taxonomy of interdisciplinarity. In R. Frodeman, J. T. Klein, and C. Mitcham (eds). *The Oxford handbook of interdisciplinarity*. Oxford: Oxford University Press.

Klenk, N. & Meehan, K. (2015). Climate change and transdisciplinary science: Problematizing the integration imperative. *Environmental Science & Policy*, 54, 160–167.

Mauser, W., Klepper, G., Rice, M., Schmalzbauer, B. S., Hackmann, H., Leemans, R., & Moore, H. (2013). Transdisciplinary global change research: The co-creation of knowledge for sustainability. *Current Opinion in Environmental Sustainability*, 5, 420–431.

Molle, F., Floch, P., Promphakping, B., & Blake, D. J. H. (2009). The "Greening of Issan": Politics, ideology and irrigation development in the Northeast of Thailand. In F. Molle, T. Foran & M. Käkönen (eds). *Contested waterscapes in the Mekong Region: Hydropower, livelihoods and governance*. London, Sterling, and VA: Earthscan, 253–282.

Mohtar, R. H. & Lawford, R. (2016). Present and future of the water-energy-food nexus and the role of the community of practice. *Journal of Environmental Studies and Sciences*, 6(1), 192–199.

Newell, B., Marsh, D. M., & Sharma, D. (2011). Enhancing the resilience of the Australian National Electricity Market: taking a systems approach in Policy Development. *Ecology and Society*, 16(2), 15. www.ecologyandsociety.org/vol16/iss2/art15/.

Osborne, P. (2015). Problematizing disciplinarity, transdisciplinary problematics. *Theory, Culture and Society*, 35, 3–35.

Polk, L. (2014). Achieving the promise of transdisciplinarity: A critical exploration of the relationship between transdisciplinary research and societal problem solving. *Sustain. Sci.*, 9(4), 439–451.

Pratt, M. L. (1991). Arts of the contact zone. *Profession*, 33–40.

Rasul, G. (2014). Food, water, and energy security in South Asia: A nexus perspective from the Hindu Kush Himalayan region. *Environmental Science & Policy*, 39(0), 35–48.

Schwanen, T. (2018). Thinking complex interconnections: Transition, nexus and geography. *Transactions of the Institute of British Geographers*, 43(2), 262–283.

Scoones, I., Leach, M., & Newell, P. (eds) (2015). *The politics of green transformations*. London: Routledge.

Sharmina, M., Hoolohan, C., Bows-Larkin, A., Burgess, P. J., Colwill, J., Gilbert, P., Howard, D., Knox, J., & Anderson, K. (2016). A nexus perspective on competing land demands: Wider lessons from a UK policy case study. *Environ. Sci. Policy*, 59, 74–84.

Stengers, I. (2011). *"Another science is possible!" A plea for slow science*. Brussels: Vrije Universiteit. Available at http://we.vub.ac.be/aphy/sites/default/files/stengers2011_plea slowscience.pdf (accessed 10 May 2017).

Stirling, A. (2015). *Developing 'nexus capabilities': Towards transdisciplinary methodologies*. The Nexus Network. Available at www.thenexusnetwork.org/new-discussion-paper-on-tra nsdisciplinary-nexus-methods-from-andy-stirling/ (accessed 01 March 2018).

Thompson, M. (2008). *Organising and disorganising: A dynamic and non-linear theory of institutional emergence and its implications*. Axminster, UK: Triarchy Press.

Verweij, M. & Thompson, M. (eds) (2006). *Clumsy solutions for a complex world*. Basingstoke, UK: Palgrave/Macmillan.

Thoren, H. (2015). The Hammer and the nail: Interdisciplinarity and problem solving in sustainability science. PhD Thesis. Lund, Sweden: Lund University, Department of Philosophy.

Wichelns, D. (2017). The water-energy-food nexus: Is the increasing attention warranted, from either a research or policy perspective?. *Environmental Science & Policy*, 69, 113–123.

Zhang, X. & Vesselinov, V. (2016). Energy-water nexus: Balancing tradeoffs between to-level decision makers. *Appl. Energy*, 183, 77–87.

5

HYBRID GOVERNANCE AND GROUNDING THE NEXUS

Nexus and integrated management are on-going, unresolved problems of complex development and its governance. Its cross-sector connections and cross-border impacts pose a major challenge to policymakers. Some scholars have argued that the main problem is not the scarcity of a resource or the lack of solutions, but the lack of political will to implement integrated long-term measures for managing resources and risks sustainably (Beisheim 2013: 2). For others, the nexus literature does not clearly explain why the barriers to achieve coherence are present, what influences them, and how they can be acted upon. In particular, it falls short on providing insights on (i) conditions for cross-sector coordination and collaboration; (ii) dynamics that influence the nexus beyond cross-sector interactions; and (iii) political and cognitive factors as determinants of policy change (Weitz et al 2017: 165).

The nexus is yet to be extensively grounded into national policies and practices, and broad-based local demand for nexus-framed policies is currently limited. The opposite of a more holistic or interdisciplinary water–food–energy nexus approach is silo-fication, which is the consequence of hierarchic organising and specialisation at levels of social organisation above (the primary one) the farming family.

This chapter will not necessarily focus on the hegemonic official nexus policy and its impact on the ground, but rather look at how nexus issues are being constructed under hybrid governance systems involving local institutions and individuals. Through a series of mini-case studies in Nepal, this chapter will seek to analyse what is the local understanding and practice around the relationship between food, energy, and water to inform nexus thinking and practice. These mini-case studies inform us on the interaction between formal and informal institutional arrangement and how these interactions form the basis of a nexus system.

This chapter's contention is that, given the lack of success of previous efforts at "integrated management", there is a need to ask how the water policy terrain can be pluralised, and how space can thereby be provided to multiple voices who define very differently what the water problem is. Nexus thinking can be forwarded either by fortuitous enlightened statesmanship or disasters, both of which can be taken advantage of if they arise but neither of which can be planned or wished for. In normal mundane times, alternatively, one should strive to break silo-thinking by letting plural voices be both heard and responded to; and the very process of that constructive engagement between different social styles of organising – hierarchism, individualism, and egalitarianism – would structurally be in itself the integrating factor between silos nexused at a more basic level.

5.1. Whither nexus governance?

There are fundamental physical attributes and differences between the water, energy, and food sectors that must be taken into account when considering each other's limits, synergies, and entwined predicaments. Energy can be produced from many sources and in different locales: water cannot be manufactured and one has to live within its limits. Energy – especially non-renewable – can be mined and the institutional requirements for it can be handled as a security issue: just send in your army, secure the site, and mine away (or "drill baby, drill"). Water (and food) needs to be primarily harvested and it is only at their specialised ends that they need specialised processing.

It is argued here that, if we are to achieve a more nexus-like approach of trade-offs between the water, energy, and food sectors, if we are to see a move from unstable monistic or even dualistic governance therein to one where varied voices of different social solidarities with differing perceptions of risk and technology choices co-exist, then different definitions of "what the problem is" must find a way to the policy table and must be responded to by other voices. The normal trend is towards bureaucratic silo-fication on the one hand, and a giving up of all regulation to market individualism on the other. This monism or the only slightly more plural "public–private partnership" is still deficient in that it ignores the critical voices of egalitarianism that often have the same function as canaries in a mine. Despite a few false starts and often excessive alarmism, they do have a critical function in preventing (or at least anticipating) unpleasant surprises as happens with "boil-over" in societies that the water–food–energy crises of 2008 demonstrated.

Nexus thinking has evolved from the development graveyards of integrated development efforts of the 1970s and 1980s. It is important to understand and reflect upon why integration did not occur, why nexus may similarly fail in monistic or dualistic policy terrains, and why non-market and non-hierarchic values and approaches must also find space in governance and policy framing. The contention of Cultural Theory (or the Theory of Plural Rationalities, see for example Beck et al 2018; Douglas 1987; Thompson 2008) is that, when integration or nexus are approached as too much of a hierarchic regulatory issue, it is bound to fail if the concerns of market individualism or activist justice-seeking egalitarianism are not met. Given that these primary social solidarities exist at all levels from village to global commons, it is also contended that the closer social organising is to that of the primary social organising form, i.e. the family, the easier for nexus approach to find acceptance. And, as it moves further away, from village councils to municipalities to district headquarters to national and international agencies, more entrenched silo-fication takes place. Such a move towards specialisation and proceduralism needs to be counterbalanced by oversight bodies that cross-cut across departmental boundaries.

A key question is what are the innovations needed for a nexus approach to solving intertwined problems and where will they come from? Again, the understanding of Cultural Theory (see for example Gyawali 2009; Verweij & Thompson 2006) is that innovations are needed in all three of the different styles of organising and at each level of organising as per the inherent strengths of each. These three styles of organising deploy different types of power: government bureaucracies often apply coercive power ("follow the law or go to jail"); markets deploy persuasive power ("buy this product and feel yourself in heaven"); and civic movements specialize in moral power ("you are being bad to society, environment, the poor ... when you ..."). It is with these tools that each of the different styles of organising has to engage with the others in a nexus approach and to concentrate at points where they perforce intersect.

Some points of intersection have been discussed more than others in the nexus approach to governance. For example, both storage and transportation are two intersections that require more debate, development, and innovations sought. Moreover, there are two governance indicators at these intersection points where engagement has to be sought between the three active social solidarities: considerations of efficiency that markets and governments readily understand; and considerations of each other's "footprints" that civic movements and governments understand but markets do not so readily. For example, it has been estimated that 30–40 per cent of food crops are lost between production and the market globally.[1] This translates into the wastage of almost a quarter of all freshwater, crop area, and fertilizer currently used for food production! Moreover, while much of the food wasted in developed countries is at the household level, in developing countries it is at the level of field and transportation. Efficiency that markets inherently advocate can achieve nexus gains if governments frame rules of punishment, and this can only be lauded by the civic voices of the ethics community.

Water footprints or energy footprints of food as well as other consumer products are another indicator that has found favour among the civic egalitarian voices that has a strong ethical ring to it. For example, there is something inherently understandable as unsustainable and wrong if it takes far more energy to transport and store fruits and other perishable products than is there in the product itself (Khan et al 2009). It brings back into governance debates the issue of behavioural changes: what is justified need and what constitutes luxury or conspicuous consumption that should be discouraged through taxation and redistribution?

Thus, storage and transport interstices under considerations of efficiency and footprints are areas where market individualism has to bring forth technological innovations, bureaucratic hierarchism has to bring regulatory innovations, and activist egalitarianism has to bring about innovations in behaviour and value changes. It would suggest that a nexus approach could emerge without having to wait for a statesman messiah or a major disaster.

5.2. Governing the nexus as knowledge politics

Some of the key aspects of governance and decision-making are evidence and knowledge. One important approach that is emerging from these policy dynamics has been the development of nexus assessments and toolkits to assess and prioritise different development options. As Hoff (2011: 12) points out, "there is a need for a coordinated and harmonized nexus knowledge-base and database indicators and metrics that cover all relevant spatial and temporal scales and planning horizons". These models are currently being developed by many international organisations from the Food and Agriculture Organization (FAO) (Mohtar & Daher 2012), to the Asian Development Bank (ADB) (ADB 2013) and initial nexus assessments have been undertaken in various case studies around the world, e.g. in Burkina Faso (Hermann et al 2012), Mauritius (Welsch et al 2014), and the US (DoE 2014).

While water managers in their silos make decisions of a highly specialised nature, and that are very necessary for the proper functioning of complex water delivery systems such as dams and waste treatment plants, failure to appreciate or connect with the happenings "outside the water box" (essentially externalities beyond their direct control) could spell disaster for happenings within the water box and catch these water managers by unpleasant surprise. A UN report (FAO 2014) argues that decisions on how to intervene made without cross-sectoral coordination and targeting only sector-specific optima risk increasing uncertainties across sectors and scales, altering the conditions under which they were designed. They state: "In order to ensure the optimal management of trade-offs and the maximization of overall benefits, decision-making processes need to be reflective and take into account the dynamic nature of complex systems" (FAO 2014: 5).

Analytical frameworks that seek to understand trade-offs and synergies are being developed as a way to address the governance between food, water, and energy systems (e.g. Hoff 2011; Smajgl & Ward 2013). The nexus thus seeks to integrate sectors through making them visible and thereby aims to address externalities that link sectors together. It raises a need to openly discuss trade-offs and the decisions that this entails; the governance of these decisions, namely who takes decisions and for whom, is also important yet has been less rigorously discussed to date.

How knowledge politics shapes governance of the nexus is revealed in the different framings of nexus dynamics and trade-offs across different social groups/policy actors. Such an analytical approach has been developed by Kalberg et al (2015) through a case study of Lake Tana in Ethiopia. Recognising the inherent subjectivities and dynamics in policy decisions, the model developed is based on participatory scenario modelling in order to illustrate the different views and preferences of actors across different scales and sectors. The nexus toolkit developed by Kalberg et al (2015) helps to illustrate system-wide and cross-sectoral outcomes of different policies, interventions, and development trajectories, allows an assessment of the potential synergies of a nexus approach, and identifies any remaining dilemmas that need to be agreed upon by actors. More importantly, the approach can be used as mutual learning platforms and processes that help to understand the assumptions and expectations of the various actors involved in the policy problem. One of the key questions in relation on this approach, which is becoming mainstream, is the number of assumptions both behind the scenarios and their translation for policy relevance. In this particular case, the framing of the nexus was enlarged to integrate the ecosystem dimensions and the current water level of Lake Tana was taken as a major dilemma to be addressed in current policy framework if water use remains unregulated for both irrigation and hydropower production purposes. Kalberg et al (2015; see also Benson, Gain, & Rouillard 2015) are thereby transposing the nexus logic to national and subnational frameworks. In this respect, the nexus is very much seen as a state centric logic of maximising national resources.

An article by Villamayor-Tomas et al (2015) applies a value chain analysis, which assumes that there is some rational knowledge-based, potentially optimisable way of allocating and managing water and energy. He argues that the private sector supply chains have largely determined energy and climate change policies. The question is how these can operate at the nexus level between water, food, and energy. They illustrate in different case studies how food and energy market systems do not integrate a number of issues for a value chain analysis of the nexus. Examples are given and discussed, such as the economic and environmental risks in food supply chains of consuming vast volumes of unaccounted water associated with inadequately costed embedded energy incurred in pumping water; or the consequences of wasteful allocation and management of water and energy in the production, conveyance, and (wasteful) consumption of goods and services in which the real but unaccounted costs of water and energy are substantial. They argue that profits have not proved to be a sound metric to incentivise reliable stewardship of natural ecosystems, even as it is the main incentive for the market players who deliver water intensive food and energy services. Strong regulations and adequate institutions are therefore needed to create these incentives. In fact, most scholarly and policy document on the nexus fail to consider the role of institutions in mediating environmental outcomes. As argued by Villamayor-Tomas et al (2015), nexus frameworks are process-based approaches to resource use and show a preference for systems analysis and modelling over other empirical strategies. This can be useful to study water–food–energy inter-linkages, but neglects the very important role of institutionally-mediated human agency. One of the most interesting conclusions of this article is the resistance to change. The coupling of water and energy use has increased over time and previously independent institutions need to adapt to take into account cross-sector linkages. However, as illustrated in the Indian and Spanish cases in their article, vested interests and norms may prevent such adaptation and reflect institutional path dependencies. Once you move along the de-nexused, silo-fied path of heavy duty engineering, you create technological lock-ins and lose flexibility and it becomes too costly to change to an alternative pathway (Beck et al 2018).

This dichotomy between synergy and conflict, as is often described in the nexus literature, is far too simplistic. Guiding principles are essential when negotiating over diverging interests. Although some overarching nexus principles have been proposed in the nexus literature (e.g. investing to sustain ecosystem services, creating more with less, and accelerating access by integrating the poorest (Hoff 2011), there is no consensus on them (Weitz et al 2017: 168).

Overall, the current problem with these approaches is that they rarely take into account the current social formations that energy, food, and water systems have produced historically (Foran 2015). If uncertainty and complexity are indeed to be taken seriously, the links between cause and effect, between premise and conclusion, cannot be isolated to proximal cause and effect, but should also include circumstantial premises, for example, statements about historical and social conditions that when taken together, help justify arguments about causality (Fairclough & Fairclough 2012).

5.3. Grounding the nexus: Reflecting on the tug-of-war between "organising–disorganising" forces at work

5.3.1. Case 1: Kulekhani Hydroelectric Power Stations

The case of Nepal's only large reservoir, the Kulekhani, is instructive in understanding how, despite the multiple benefits of a nexus approach, the ground imperatives of government agency practices promote de-nexusing the water–food–energy sectors that converge around the reservoir into a siloed (or also sometimes called stove-pipe) approaches.

The cascade of Kulekhani Hydroelectric Power Stations is located 30 km southeast from Kathmandu in Dhorsing, Village Development Committee (VDC) of Makawanpur district. Kulekhani is the only seasonal storage type hydroelectric project in the integrated Nepali power grid and the only large dam the county has. Two power stations already operational within this cascade complex are the 60 MW Kulekhani No 1 (KL-1) at the upper end of the cascade that has a high dam of 114m and a reservoir. When the KL-1 was completed on 14th May 1982 it was the largest generating unit in Nepal. It stores a gross volume of 85 million cubic meters (MCM) and a live volume of 73 MCM of monsoon flow in the eponymous river which is not snow-fed. A lower 32 MW Kulekhani No 2 (KL-2), completed in November 1986, uses the tailrace waters of KL-1 and a 14 MW KL-3 is nearing completion further below, despite severe time and cost overruns, which will use the tailrace of KL-2.

Conceived purely as a peaking hydroelectric plant (KL-1), considerations of using the stored water either within the reservoir for fisheries or downstream for irrigation and drinking water were never part of the official project design by the government agencies nor proposed as such by the World Bank as well as the

Japanese aid agency JICA bankrolling it. There have been writings by activists and academics in the local media suggesting nexus activities. They range from: using the higher water level in the reservoir to supply gravity flow drinking water to the chronically water scarce capital city of Kathmandu located at a lower altitude; promoting tourism and fisheries in the lake; increasing dry season irrigation in the downstream reaches from the stored releases; providing more municipal water supply to the town of Hetauda below KL-3; and enhancing environmental flows to the national wildlife parks in the downstream reaches. While obviously bene-ficial, they have fallen on deaf official ears that refuse to see the reservoir as any-thing other than a peaking pond for hydroelectric generation. It has, however, not prevented other social solidarities (informal markets as well as social and environ-mental groups) from exploiting the created resources or advocating for the same. This disjuncture between the official and the unofficial, the formal versus the informal, and the national versus the local is what this case study strives to high-light. Unlike other storage dams in Nepal, Kulekhani is on a small fourth order tributary of the Ganges without an international transboundary element in its debates, which is purely focused on domestic Nepali issues and which should, in principle, have been easier to resolve.

KL-1 was originally investigated by His Majesty's Government of Nepal in the 1970s with assistance from the Japanese JICA as part of the Fifth Five-Year Plan, which outlined the objectives of power development as: (1) to further develop hydroelectric resources to meet increasing power demand; (2) to extend electric services gradually to new areas in the country; and (3) to bring about regional bal-ance in the production and distribution of electric power. KL-1 actually defeated the third objective as it helped make the central region of the country totally dominant in the integrated power system of Nepal. This project was financed through a combination of various foreign sources, in addition to the government of Nepal, including the World Bank, the Kuwait Fund, the United Nations Development Programme (UNDP), the Overseas Economic Cooperation Fund (OECF) Japan, and the Organisation of Petroleum Exporting Countries (OPEC).

When the project was seriously considered for implementation in the early 1970s, there was virtually no opposition of any kind. The relatively small number of people who were to be displaced by the dam were just given the official com-pensation and told to move away. There is a poor record of who exactly were displaced and what they received. In interviews with people in the catchment area, it is mentioned that those displaced moved on downstream to the townships of Hetauda or elsewhere in search of livelihood, most having misused or misspent their compensation amounts.

The population surrounding the Kulekhani Watershed, and most of rural Nepal, has a deep and integrated relationship with the geography and land where liveli-hoods and culture depend on this relationship with the environment. Resilience and adaptability is a constant trend living in the middle hills of Nepal. With their susceptibility to the effects of climate change and major developments like the Kulekhani hydroelectric project (HEP), villagers have continued to live and make use of the land despite the changes to the ecosystems as a result of the construction of the dams, creation of the reservoir, and the socio-economic effects of the process of major infrastructural development.

The creation of the reservoir also brought about (as an afterthought) the potential for new livelihood options, including aquaculture and improved fishing opportunities, and the use of dry season regulated/stored water for downstream municipal uses. Studies were completed following the creation of the reservoir to determine the aquaculture and fishing potential of the new reservoir. It was found that indigenous species of katale (*A. hexagonolepis*) and karange (*P. chilinoides*) adapted to the conditions present in the reservoir, while asala (*S. richardsoni*) all but vanished in this area. Three additional species, silver carp (*H. molitrix*); bighead carp (*A. nobilis*); and mahseer (*T. tor*), were introduced as part of aquaculture programs (Pradhan & Swar 1988). These introduced species were ultimately hailed as a general success in creating livelihoods and diversification of diets of the local population, but the introduced species were also seen as further altering existing ecosystems. New livelihood opportunities in aquaculture have been hampered by disjunctured government utility maintenance problems resulting in fluctuating oxygen levels and changes in ecosystem conditions as a result of continued hydropower development and construction.

The real conflicts with the project began long after its completion with the restoration of multiparty democracy and the ability of the populace to voice public grievances. In July 1993, a major disaster struck the project when an intense cloudburst, lasting 30 hours with intensity of up to 60mm/hr, dumped as much as 540 mm of rain in just 24 hours. There was much mass wasting and landslides in the catchment area that practically filled up KL-1's entire dead storage volume that was planned to last 100 years. Bridges and sections of the national highways were washed away as were 67 small and large irrigation projects; and some two thousand people lost their lives. Subsequent bathymetric surveys indicated that actual sedimentation was orders of magnitude higher than designed for. The torrential rains dislodged hill slopes and washed away the penstock of KL-1 shutting down its operation completely (equivalent to 40 per cent of the total grid power), necessitating serious (and expensive) counter measure constructions. An innovative "sloping intake" was constructed that allowed the intake point of the headrace tunnel to move up as the lower part of the reservoir fills up with sediment. It was during this phase of rehabilitation/reconstruction that conflicts came to the fore highlighting the nexused nature of the reservoir.

The people who lived within the catchment around the reservoir area but who lost their lands at the valley bottom when the river was dammed began cage fish farming with encouragement by activists and some Japanese volunteers. There were no official agreements with the national utility that managed the dam; but the officials were not bothered either since it did not affect their power generation. When the 1993 disaster struck and the sloping intake had to be constructed to make the plant functional again, the utility resorted to a sudden and quick dewatering of the reservoir killing all the fish that the villagers had been farming. A massive conflict ensued at the local level.

The initial official utility position was that it was their pond and they could do what they liked, "even pour poison in the reservoir", and that the people fishing there had no official right to do so. Given that multiparty democracy had just been restored in Nepal, it was difficult for the political parties to go along with the hydrocracy's position, and a compromise of sorts was worked out. The fisher folks would be paid a one-time compensation of almost a million rupees and they would be free to continue with their fish farming in an informal way. However, if anything untoward happened due to reservoir operations by the utility, they could not claim any compensation in the future.

The fish farming continues de-nexused in the informal economy engaging some 307 families around the reservoir area. They are now self-organising into a self-help cooperative with members having fixed shares so as to prevent overfishing through self-regulation. There is virtually no national government regulation, and local elections were held after a gap of over twenty years but the local bodies are still not functional (at the time of writing) due to lack of appropriate enabling legislation. Cage fishing has now declined in favour of capture fishing, which constitutes about 80 per cent of the catch, with aquaculture for the remaining 20 per cent, mainly in the cold winter season. In 2013, the families sold some 52 thousand kgs of fish. The as-yet unregistered cooperative also allows tourists to do some angling for a fee of 500 rupees, provided that they catch less than 5kgs; if the catch more they have to buy it back from the cooperative.

The catch amount varies. It is about 100–150 kg per day in the winter, 200–300 kg during the monsoon and up to 500 kg in the post monsoon months for all the 307 families. Given that Nepal government rules now stipulate that some 12 (very local) plus 38 (district in the region) (equalling 50) per cent of the government's royalty from hydropower plants, have to go to local village and district governments (in a complex and as yet not streamlined – but contested – set of formulas), the fisher folks are demanding that some of the money come to them too and that they receive help in running a motorboat service (replacing paddled dugout canoes) for transport between villages on opposite sides of the lake. All this is being done quite openly and mostly in the informal sector with practically no "nexus" with the electricity generating people below (whose families in the Dhorsing staff quarters at the powerhouse site, however, buy the fish for their private consumption!).

The other set of "de-nexused" activities is downstream of the dam. Interviews with two sets of beneficiaries from the regulated waters coming out of the tailrace of the Kulekhani system were conducted: the informal farmer-managed irrigation systems (FMIS) along the Rapti River where the tailrace empties into, as well as the Hetauda municipality. There are several what are called traditional (and in the informal system) FMISs in the vicinity that have used the waters of the Rapti River for generations, and who now use the extra releases from Kulekhani HEP operations in the dry season. These irrigation systems range in size from 35 hectares to 150 hectares and use traditional technologies that divert the river waters for irrigation with brushwood dams and simple gravity flow unlined canals. In very dry seasons, diesel and electric pumps are sometimes used to pump supplemental water from the river itself.

Exact quantification is near impossible, but it is obvious from interviews with farmers that the released waters from the tailrace of KL-3 have increased the flow in the river, especially in the very dry seasons of February to May, allowing for more irrigation than would have been possible without the upstream reservoir. The official irrigation department of the government is only now beginning to look into this phenomenon, and that too (when the interview was conducted) because a new Department of Irrigation official appointed to its Hetaunda office happened to be a local resident.

The other "de-nexused" player is the industrial township of Hetauda Municipality and its water supply system on the Rapti River downstream of the KL-3 tailrace. This town is experiencing growing water shortage and is eyeing the stored waters released by the Kulekhani HEP tailrace to augment its supplies. Initially, this matter was not considered as an option (it was filtered out) while the town's water supply was operated by the national level Water Supply and Sewerage Corporation managed out of the capital Kathmandu (which manages the water and sewerage systems for all major towns of Nepal). However, due to increased public pressure recently, the water supply system has been decentralised to a local board of Hetauda residents. With this bit of devolution of power to a lower unit of government, suddenly nexus thinking has arrived and consideration of augmenting the town's water supply from the KL-3 tailrace are very much on the table.

There are, however, many negotiations ahead between these four very different players of electricity, irrigation, fisheries, and domestic/industrial water supply – a clear case of de-nexused development and a re-nexusing dynamics underway. Thus, a de-nexused water–food–energy nexus exists in the largest and only big reservoir in Nepal; and the nexusing is happening only with local and informal initiatives, but not at the official national Nepal government or international aid agency levels.

5.3.2. Case 2: Melamchi transbasin water supply

That Kathmandu Valley, housing Nepal's capital, has been facing chronic water shortage is a truism easily believed, especially by its denizens. Given that the Valley with a catchment area of about 600 km2 receives 1200 mm of annual rainfall, it stands to reason that natural shortage could not be that much of a cause of the scarcity. "Unmetered consumption" is rife, as are technical losses from leaky pipes and tanks. However, water scarcity is the essence of a local political narrative; and since the 1970s, Melamchi transbasin transfer of water to the Valley has been the iconic project of salvation. Diverting water from the upper reaches of the Melamchi River, a tributary of the snow-fed Sun Kosi north of Kathmandu into the valley was identified as a viable measure to remedy the water shortage problem of Kathmandu valley some forty years ago, in the early 1970s. The World Bank picked it up as its flagship project in the 1980s but had to pull out of it in the mid-1990s, primarily because the conditions of the Bank attached to the funding were not politically and administratively palatable to the government and partly because adding more supplies to a system that leaked (or suffered pilferage of) more than half its waters was hardly justifiable (Gyawali 2015).

After the World Bank pulled out, the ADB jumped in to promote this project, initially in partnership with the Norwegian aid agency. However, some fall-out occurred among these development partners, the Norwegians being interested in a hydropower component which the ADB and the Nepal government Ministry of Housing and Physical Planning were not interesting in adding as a complication. It is now an ADB-led venture (with support from Japanese JICA and the OPEC Fund) with only the water supply component on its horizon. The Nepali line agency for this project was the Department of Water Supply and Sewerage of the Ministry of Housing and Physical Planning, whose mandate was water supply and within which there was strong resistance to adding electricity or any other component that would attract the quest for a project share by other line agencies. Thus, the stage was set for this very large project for Nepal becoming a single-purpose water supply project for Kathmandu.

According to projections made in an early conceptualisation of this project by the Nepal Water Supply Corporation, Kathmandu valley would have a demand of 310 million litres per day (MLPD) of water in 2010 increasing thereafter by about ten percent per annum. However, the Melamchi project will be adding only about 170 MLPD during the dry season to the existing volume of supply which is 90 MLPD. This volume is insufficient in view of the projected demand that requires another 100 MLPD. Furthermore, taking the officially acknowledged leakage level of 40 per cent into consideration (other studies suggest 70 per cent), the actual amount of water available in Kathmandu valley will be mere 156 MLPD even after augmenting it with the water diverted from Melamchi. Therefore, even after the completion of Melamchi project as it is designed, the problem of water scarcity in Kathmandu valley will still persist. As ADB has forced a covenant in the loan documents requiring a massive escalation of water tariff, the valley population will face escalated prices without any reprieve in the water shortage problem.

The project involves the construction of a tunnel located at an altitude of 1700 meters on the upper reaches of the Melamchi river to bring water down to Sundarijal in the Kathmandu valley (1400 meters altitude) thereby creating a head of 300 meters so that about 25 MW power could be generated. But the idea of generating hydropower was later abandoned claiming that it was not feasible. It is true that having to dig a 27-kilometre long tunnel just to generate about 25 MW of electricity sounds too expensive even at a cursory glance. But the fact was filtered out that the tunnel has to be constructed anyway to divert water into the Kathmandu valley. Therefore, if one were to generate power from the same water, the tunnel cost is as good as free for the power generation component, and the incremental cost will be only for the construction of the powerhouse, procurement of electro-mechanical equipment, and erection thereof. This fact was completely ignored in deciding to abandon the hydropower component.

Interestingly, in contrast to the image of NGOs as obstructionist and only opposed to dams, Nepali NGOs, like most of their Southern counterparts, do not say "No Dams!" but instead argue "No Bad Dams!" In the case of Melamchi, they argue that if a river is to be sacrificed for the good of human society, sacrifice it well to maximum benefit. Instead of the current "de-nexused" single purpose design that only produces expensive drinking water (and insufficient amounts at that!) and ignores the potential for hydropower, they want the project designed for a "bigger Melamchi, multipurpose Melamchi!" They want a few other tributaries of Melamchi such as the Yangri, Larke, and Balephi also diverted to the Melamchi tunnel which should be increased in diameter from the current 3.7 meters to five meters at only incremental costs. This will bring 1120 MLPD of water to the valley in the dry season, of which only 20 per cent will be lost in consumption and evaporation while 80 per cent can be recovered with proper sewerage treatment for a healthy and clean Bagmati river (Gyawali 2015).

Unlike the current project, this "bigger Melamchi, multipurpose Melamchi" approach of the NGOs is a more nexused one that would simultaneously provide the following additional benefits. The additional diversions from the Melamchi tributaries would not only provide some 50 MW of hydroelectricity before the water enters the main Melamchi tunnel but also triple the dry season flow from six to thirteen cubic meters per second, meeting all possible future demand for drinking water in the valley. Below the Chobhar gorge at the southern end of Kathmandu valley, the head difference between it and the Tarai plains of some 900 meters could be used to generate an additional 190 MW of power for the Nepal power grid that currently suffers up to 15 hours of power cuts per day. In addition, the increased dry season flow in the Bagmati River could increase the potential irrigable command area of the lower Bagmati plains in Sarlahi and Rautahat Tarai districts by 30,000 hectares.

With this nexused, multi-purpose approach, unlike the current siloed one of treating Melamchi as a drinking water project only, Kathmandu Valley would get more reliable and cheaper drinking water since electricity users across the country would pay for part of the cost of the tunnel. With more flow in the river as well as sewerage treatment plants, the Bagmati River running through the capital city would be cleaner, unlike the foul sewer it has currently become. The national grid would get much needed hydroelectricity to ameliorate its severe load-shedding; and the Tarai plains would get dry season irrigation to increase food production. Unfortunately, with the current silo approach, only Kathmandu gets expensive and insufficient water; Bagmati River remains a sewer; Tarai gets no irrigation, only Kathmandu's untreated effluents; and the country gets no electricity! And this example additionally shows that silo-ed, de-nexused thinking is not only the bane of developing countries but also international development agencies.

5.4. From IWRM to nexus: The challenges of siloisation?

As mentioned earlier, such de-nexused management pathologies can be found in various other places in the Global South. In a comprehensive look at water, energy, and food issues in India, for example, Kumar et al (2014) examine the intertwined nature of these three sectors in what is essentially a semi-arid region with poorly endowed energy resources and heavy population pressure on its food-producing land resources. Each of these sectors is also plagued by the "securitisation" approach as well that of "populism" in policy making wherein a dominant alliance of politicians–bureaucrats–academics is pushing for short, quick-fix solutions to the long-term detriment of the economy of these sectors. Absence or lack of appreciation of the nexused nature of these intertwined sectors, together with issues of rural labour scarcity as well as populist underpricing of electricity for groundwater pumping, socially insensitive promotion of hydropower construction, and cartel/monopoly ownership of groundwater wells is contributing to growing insecurities in each of the sectors.

What this and other cases demonstrate is that the reasons why IWRM did not succeed in getting traction on the ground are still alive and strong against the nexus approach (see Chapter 3). To continue the Kulekhani story from Nepal described above, the country had a Ministry of Water Resources (MoWR) that was responsible for hydropower (Nepal's official national grid is mostly hydroelectric), irrigation, flood control, and drinking water under its jurisdiction. The last was hived off to the "housing ministry" in the late 1980s. In 2010, the Water Resources ministry was split off into a Ministry of Energy and another Ministry of Irrigation, thus de-nexusing even further the silo-ised departments under their jurisdiction. This "dis-integation" happened despite the fact that the country's official water policy had enshrined the IWRM principle within it, and despite the fact that the majority of the bureaucracy was opposed to splitting MoWR (although a small minority that ultimately prevailed wanted to carve out fiefdoms). The prime obvious driver was the compulsions of coalition party politics where more party cadres had to be found ministerial berths than there were ministries available.

Moreover, Nepal also has a Water and Energy Commission (WECS) that has, as members, secretaries of some twelve major ministries such as water, power, petroleum supplies, agriculture, forests, finance, foreign affairs, etc. In principle this institutionally nexused body could serve as the point of nexus thinking and policy trade-offs. In practice, except in the 1980s, this body has been used mostly as a bureaucratic "shunting yard" where senior bureaucrats out of favour with the ruling dispensation at the time are shunted off to cool their heels. Major decisions on large water resources treaties (i.e. the 1996 Mahakali Treaty with India) or projects (Arun-3 led by the World Bank from which the Bank pulled out after protests in 1995) were either never taken to the WECS for advice or if taken, and the advice was unfavourable, was ignored.

While decisions on the water, energy, and food sectors may be nexused at the level of the farming household, the process of silo-fication gains prominence and strength once we move to higher levels of governance. Similar to university departments where "interdisciplinary studies" are often prominently lip-serviced but actual academic promotion depends very much on disciplinary contributions, rewards in a siloed structure of bureaucracy is via promoting siloed interests and "empire building" for the concerned silo.

Note

1 www.fao.org/in-action/seeking-end-to-loss-and-waste-of-food-along-production-chain/
en/ (accessed 10 June 2018).

References

ADB (2013). *Thinking about water differently: Managing the water-food-energy nexus*. Philippines:
Asian Development Bank.

Beck, M. B., Thompson, M., Gyawali, D., Langan, S., & Linnerooth-Bayer, J. (2018).
Viewpoint–Pouring money down the drain: Can we break the habit by reconceiving
wastes as resources? *Water Alternatives*, 11(2), 260–283.

Beisheim, M. (2013). The water, energy & food security nexus: How to govern complex
risks to sustainable supply. *SWP Comments*, 32, 1–8.

Benson, D., Gain, A. K., & Rouillard, J. J. (2015). Water governance in a comparative
perspective: From IWRM to a 'nexus' approach? *Water Alternatives*, 8(1), 756–773.

DoE (2014). *The water-energy nexus: Challenges and opportunities*. Washington, DC: Department of Energy.

Douglas, M. (1987). *How institutions think*. London: Routledge and Keegan Paul.

Fairclough, I. & Fairclough, N. (2012). *Political discourse analysis: A method for advanced students*. London: Taylor & Francis.

FAO (Food and Agriculture Organization of the United Nations) (2014). *The water-energy-
food nexus: A new approach in support of food security and sustainable agriculture*. Rome: FAO.

Foran, T. (2015). Node and regime: Interdisciplinary analysis of water-energy-food nexus in
the Mekong region. *Water Alternatives*, 8(1), 655–674.

Gyawali, D. (2015). *Nexus governance: Harnessing contending forces at work, nexus dialogue
synthesis papers*. Gland, Switzerland: IUCN.

Gyawali, D. (2009). Pluralized water policy terrain: Sustainability and integrati on. Viewpoint in eJournal *SAWAS*. Hyderabad: South Asian Water Studies.

Hermann, S., Welsch, M., Segerström, R. E., Howells, M., Young, C., Alfstad, T., Rogner,
H. H., & Steduto, P. (2012). Climate, land, energy and water (CLEW) interlinkages in
Burkina Faso: An analysis of agricultural intensification and bioenergy production. *Natural
Resources Forum*, 36, 245–262.

Hoff, H. (2011). Understanding the nexus. Background Paper for the Bonn 2011 Conference: The Water, Energy and Food Security Nexus. Stockholm: Stockholm Environment Institute.

Khan, S., Khan, M. A., Hanjra, M. A., & Mu, J. (2009). Pathways to reduce the environmental
footprints of water and energy inputs in food production. *Food Policy*, 34(2), 141–149.

Kalberg, L., Hoff, H., Amsalu, T., Andersson, K., Binnington, T., Flores-López, F., de
Bruin, A., Gebrehiwot, S. G., Gedif, B., zur Heide, F., Johnson, O., Osbeck, M., &
Young, C. (2015). Tackling complexity: Understanding the food-energy-environment
nexus in Ethiopia's Lake Tana sub-basin. *Water Alternatives*, 8(1), 710–734.

Kumar, D. M., Bassi, N., Narayanamoorthy, A., & Sivamohan, M. V. K. (eds) (2014). *The water,
energy and food security nexus: Lessons from India for development*. London: Routledge Earthscan.

Mohtar, R. & Daher, B. (2012). Water, energy, and food: The ultimate nexus. In *Encyclo-
pedia of Agricultural, Food, and Biological Engineering*, Second Edition. London: CRC Press.

Pradhan, B. R. & Swar, D. (1988). Limnology and fishery potential of the Indrasarobar
Reservoir at Kulekhani, Nepal. In S. S. De Silva (ed). *Reservoir fishery management and
development in Asia. Proceedings of a workshop held in Kathmandu, Nepal, 23–28 November
1987*. Kathmandu: Department of Fisheries, Ministry of Agriculture, and IDRC Canada.

Smajgl, A. & Ward, J. (eds) (2013). *The water-food-energy nexus in the Mekong Region: Assessing Development strategies considering cross-sectoral and transboundary impacts.* New York, Heidelberg, Dordrecht, and London: Springer.

Thompson, M. (2008). *Organising and disorganising.* Axminster, UK: Triarchy Press.

Verweij, M. & Thompson, M. (eds) (2006). *Clumsy solutions for a complex world.* Basingstoke, UK: Palgrave/Macmillan.

Villamayor-Tomas, S., Grundmann, P., Epstein, G., Evans, T., & Kimmich, C. (2015). The water-energy-food security nexus through the lenses of the value chain and the institutional analysis and development frameworks. *Water Alternatives,* 8(1), 735–755.

Weitz, N., Strambo, C., Kemp-Benedict, E., & Nilsson, M. (2017). Closing the governance gaps in the water-energy-food nexus: Insights from integrative governance. *Global Environmental Change,* 45, 165–173.

Welsch, M., Hermann, S., Howells, M., Rogner, H., Young, C., Ramma, I., Bazilian, M., Fischer, G., Alfstad, T., Gielen, D., Le Blanc, D., Röhrl, A., Steduto, P., & Müller, A. (2014). Adding value with CLEWS: Modelling the energy system and its interdependencies for Mauritius. *Applied Energy,* 113, 1434–1445.

6

NEXUS RIGHTS AND JUSTICE

6.1. Introduction

Issues of rights and justice have been neglected in debates about the water–food–energy nexus. Globally, the ongoing industrial-scale commodification of water, energy, and land "from above" (cf. Hall, Hirsch, & Li 2011) is producing complex and profound agrarian changes. For people living in rural places, these forces can create highly disruptive, multiple enclosures, and destabilising exclusions (Barney 2007; 2009; Bernstein 2010). Smallholders frequently experience rapid and negative impacts as various resources upon which their livelihoods depend are grabbed, diminished, or degraded. Furthermore, these impacts are cumulative as they act across multiple nexused resources and are important as farmers are also at the same time fishers, forest users, and hunter-gatherers, reflecting the often diverse livelihoods of local people – based on everyday use of multiple resources – that crucially connects aquatic and terrestrial environments (e.g. Baird & Barney 2017; Roberts 2015). The linkages and connections are generally under-recognised, by private developers, international development agencies, governments, and researchers, as this involves paying close attention to what Batterbury (2001: 441) terms the "micro-politics of livelihood decision-making", in particular places and communities. Such cumulative impacts are now shaping the character of contemporary agrarian transitions.

The combined social and environmental changes wrought by resource projects – hydropower dams, mines, industrial zones, and the like – can thus produce particular challenges for communities, as multiple systems are enclosed and degraded. Environmental and social impact assessment largely do not address the cross-sectoral and cumulative effects of such nexus-related projects (Leach, Stirling, & Scoones 2010). Where cumulative impacts have been considered, studies have focused on a single sector (such as multiple hydropower dams). A separation between water, land, and energy management frequently leads those assessing project impacts to overlook or underestimate project outcomes including cumulatively across both projects and sectors. There is often a lack of political will to seriously consider impacts across sectors. Due to ontological (Lahiri-Dutt 2014; Lavau 2013) and bureaucratic-institutional (Gober et al 2013; Middleton et al 2015) divides between water, energy, and land management, regulation and expertise, both public and private regulatory systems in many countries are currently ill equipped to effectively manage these complex cross-sectoral and multi-scaled transformations. The stakes of this disconnect are high in terms of human rights and justice, as these cumulative impacts are particularly damaging to rural livelihoods.

Globally there is also growing recognition of the relationship between the environment and human rights, including the Right to Water, as well as the role that extra territorial obligations (ETOs) might play in protecting these human rights (Boelens, Perreault, & Vos 2018; Middleton 2018). Given that human rights are interdependent and indivisible, a human rights-based approach to the water–food–energy nexus could anchor "the nexus" in a clear normative framework. That said, the right to resources (such as water, food, or energy) should not be simply understood as defining the access to the resource by the individual. Implicitly, it also relates to decision-making over who can access it, and is thus also fundamentally a social relationship and therefore necessarily also an expression of power. Thus, rights are best understood as politically contested and culturally embedded relationships among different social actors. Furthermore, beyond rights as defined in international human rights law, local rules and rights of any particular community to utilise resources emerge from place-based processes of negotiation and/or contestation that are socially-, culturally-, historically-, and politically-specific. These local rules and rights in turn reflect "power relations, local identities and contextualized constructions of legitimacy" (Boelens and Zwarteveen 2005: 735). Multiple water access and use rules commonly co-exist or are in tension, usually between national law and local customary arrangements (i.e. legal pluralism) (Roth, Boelens, & Zwarteveen 2015). Which rules are viewed as legitimate and to be applied are often contested between actors of divergent interests and power.

This chapter argues that to increase demand for nexus-thinking from below (i.e. the relevance of "the nexus" to rural communities), a technocratic ecological modernisation approach will be insufficient. Rather, the nexus concept must engage more clearly with promoting fair decision-making and just outcomes, and thus meet the expectations of many of the community resource users themselves. From the knowledge produced and deliberations held on the nexus to date, it is apparent that much of the focus has been on understanding the interaction between food, water, and energy systems, and how to shift towards sustainability including managing scarcity and trade-offs, whilst at the same time ensuring economic growth in the form of a green economy. "Good governance" of the nexus, whilst part of the nexus parlance (for example, the need for public participation), has yet to be seriously problematised and integrated, in particular in the context of the institutions, politics, and history of regions such as South and Southeast Asia. Meanwhile, justice is rarely conceptualisation or specifically addressed with regard to nexused trade-offs.

Building from Neal, Lukasiewicz, & Symec's (2014) argument that justice matters in water governance – this chapter argues that justice matters in nexus governance. Yet, defining justice in water governance from a multi-disciplinary perspective is at an early stage due to its complexity, leaving conceptualisation of justice in nexus governance at an even earlier stage. This chapter draws on environmental justice scholarship – and practice – as a promising starting point to redress this deficit. This chapter will propose that if the nexus is to support stated aspirations for sustainable development and poverty reduction, then it should engage more directly in identifying winners and losers in natural resource decision-making, the politics involved, and ultimately with the issue of justice. Thus, it will argue that introducing the concept of environmental justice (EJ) into the nexus, especially where narratives, trade-offs, and outcomes are contested, could make better use of how the nexus is framed, understood, and acted upon. To this end, it will synthesise existing EJ literature around distributional, procedural, and recognitional justice on food, water, and energy, and relate this literature to the water–food–energy nexus. The chapter highlights that social movements have explored individually injustices related to food, water and energy, but the relationship between them remain less recognized and in need of further examination. Thus, the chapter will seek to push nexus debates beyond existing calls for "good governance" to examine more critically the issues of justice at the heart of these debates, including in terms of access to and control over natural resources.

6.2. Environmental justice

The concept of environmental justice (EJ) has travelled far since its origins in the US during the 1970s. Originally a movement against the location of toxic dumping sites, chemical plants, and municipal and/or hazardous waste facilities in spaces inhabited by minority and disenfranchised populations (see Bullard 1994; Agyeman 2005), the concept of EJ has now spread to address transnational EJ often between the North and South, and within the Global South where its focus is on for example artisanal mining, children rights and mine shafts, migrant agricultural workers, and rapid, unplanned growth of human settlements and deficient infrastructure for basic services such as drinking water and sewage treatment systems (Cifuentes & Frumkin 2007; Mehta, Allouche, Nicol, & Walnycki 2014).

From its origins in the US civil rights movement (Bullard 1990), the concept of EJ has diffused around the world in the North and the South to encompass a wide diversity of local contexts together with national, transnational, and global scale concerns (Schroeder et al 2008; Sikor & Newell 2014). For example, in Southeast Asia various researchers[1] have discussed EJ in the context of large-scale water infrastructure (Middleton 2012; Simpson 2007; Sneddon & Fox 2008; Yoo 2013). More broadly, the principles of EJ are regularly claimed by civil society groups and social movements in the region, even as they may use different terminology (for example, see: Greacen & Greacen 2012; Nuntavorakarn & Sukkumnoed 2008).

In terms of its environmental boundaries, the early formulation of EJ was narrowly focused on forms of technological pollution, waste and risk – particularly those forms of "environmental bads" associated with the siting of landfill, incinerators, chemical plants, and the like. This narrowness has since given way to a far broader profile of environmental concerns (Taylor 2000) moving beyond only environmental burdens to include access to environmental benefits and resources of various forms (Mutz, Gary, & Douglas 2002) and concerns that some argue could or should be classified as social rather than environmental (Benford 2005). These shifts have extended the constituency of interested locally and nationally organised groups and increased the scope for productive interaction between environmental and social policy activists.

There is a large body of literature linking sustainable development to EJ (Agyeman, Bullard, & Evans 2003; Beder 2006; Clapp & Dauvergne 2011; Okereke 2008). Haughton (1999: 64, cited in Agyreman, Bullard, & Evans 2002), observes that finding the common ground between EJ and sustainability requires:

> acknowledging the interdependency of social justice, economic wellbeing and environmental stewardship. The social dimension is critical since the unjust society is unlikely to be sustainable in environmental or economic terms in the long run.

Many concepts of EJ readily link to those of sustainable development. Recognising the temporal consideration of sustainable development, for example, the community of justice is often understood to include the rights of future generations (Walker 2012: 10). Meanwhile regarding meeting needs that also prioritises poverty, the crux of EJ often focuses on the environmental burden and lack of access to decision-making of economically, socially, and politically marginalised communities. A particular stream of EJ, termed "environmentalism of the poor", refers to how the poor may seek to defend their existing access to resources from large-scale extraction projects, including those types of projects now associated with forms of the nexus such as large-scale land uses and large hydropower dams, because "This behaviour is consistent with their interests and with their values" (Martinez–Alier 2014: 240).

Human survivability and the sustainability of resources is already central to existing (non-global) EJ debates (Agyeman, Bullard, & Evans 2002). For many EJ activists, the link between EJ and social justice is obvious and complementary. Damaged environments damage the poorest first and foremost. Yet Agyeman (2001) points out that EJ often pays too much attention to issues of environmental quality and the implicit links between environmental improvements and social improvements while other basic human rights and equity issues and needs remain unprotected, unaddressed, and disconnected.

The concept of EJ has evolved to encompass three dimensions: distributive justice, procedural justice, and recognitional justice. Distributive aspects of justice refers to the fair distribution of both environmental risks and benefits. Procedural justice considers the ways in which decisions are made, who is involved, and who has influence[2] (Agyeman, Bullard, & Evans 2003). Justice as recognition addresses who is and isn't valued, and incorporates social and cultural (lack of) recognition, including politics of knowledge (Walker 2012:10). The presence or absence of these modes and processes of (in)justice serve to reinforce or undermine each other (Schlosberg 2004).

The concept of procedural justice, a central tenet of EJ, is embodied in Principle 10 of the Rio Declaration – a foundational document both of sustainable development and international environmental law. Principle 10 emphasises participation, access to information, and access to justice systems for redress and remedy. Procedural justice is intended to ensure that state institutions and laws and policies are fair and inclusive. Yet, EJ literature also recognises that unequal power relations together with deeper economic, social, and political structures and processes that shape the (re)production of environmental harms mean that "procedural justice" is not the only precondition to ensuring EJ. These structures and processes have been investigated by an array of social and political (ecology) theories, from common property theory and Marxist political economy to urban metabolisms and environmental governmentality (Robbins 2012: 49–81), and the subject of many such political ecology-type studies.

There is some tension, highlighted by Walker (2012: 37), between the extent that EJ and sustainability are compatible. Agyeman, Bullard, and Evans (2002: 88), reflecting on the different origins of sustainability (rooted originally in top down international policy spheres) and EJ (rooted originally in bottom up social movements) argue the discourses "have developed in parallel" and that there had been "insufficient interpenetration of values, framings, ideas and understandings". Agyeman, Bullard, and Evans (2003) subsequently argue that joined-up thinking is necessary to bring together sustainability with EJ. Some have argued that EJ movements and sustainability movements have symbiotic goals (Fisher 2003: 206). Others see the sustainability discourse as too readily acceding to the status quo of market-led development, technical eco-modernisation solutions, and power asymmetry, and thus doubt that EJ is seriously considered (Walker 2012: 37).

The principles of distribution and governance of trade-off decisions in nexus thinking – including who takes decisions and for whom – are important yet less problematised (Lele, Klousia-Marquis, & Goswami 2013). Foran (2015), for example, critiques systems approaches as under-theorised and under-politicised, in particular with regards to historical and relational considerations. He argues to link systems frameworks that identify significant nodes of interaction (Smajgl and Ward 2013) with tools of critical social science that can provide insight into the social regimes that govern those nodes.

The Stockholm Environment Institute, in a recent research brief, suggest (Davis 2014: 2):

> In some cases, however, especially when resources are very scarce, a nexus analysis may not find a win-win option, but just difficult trade-offs (Weitz et al. 2014). The role of science in such situations is not to say what the "right" answer is, but to clarify the choices and ensure that all cross-sectoral impacts, externalities and trade-offs are known and understood. Participatory processes

can also help ensure that vulnerable stakeholders have the information and access they need to advocate for themselves, and can foster dialogue across sectors and scales.

It is here that EJ literature can offer insight to the nexus. When considering trade-offs, EJ literature draws on concepts of distributional, procedural, and recognitional justice, and emphasises an explicit consideration of power asymmetries and the politics of knowledge, and other concepts such as vulnerability, needs, and responsibilities (Walker 2012: 46). Justice involves positioning of what is "fair" and thus can be highly contested and political; the notion of justice can be defined according to multiple principles, including equality of rights, utilitarian equity, and justice as fairness, as well as accommodating indigenous and customary justice arrangements.[3] Thus, in considering EJ from an analytical perspective, it is crucial to consider the framing and knowledge claims made by each actor including by policy makers, affected local communities, civil society groups, and others who may or may not share the same interests and agendas (Walker 2012: 40).

In order to critically unpack "claim-making" in EJ, recent theorisation of the EJ literature has proposed three articulated elements to be considered: theories of justice; use of evidence, including its politics; and frameworks that explain processes by which inequality and injustice is (re)produced.[4] Walker (2012: 53) states "It is the combination of evidence of inequality and claims about what makes justice and injustice that constitutes the core of the practice of environmental justice claim-making, wherever this is enacted." Evidence includes that which relates to distributional inequalities of environmental harms (or benefits), as well as procedural and recognitional inequalities. Given that knowledge is socially-produced, a politics of knowledge also exists, namely whose knowledge, evidence, or arguments is recognised as legitimate (scientists, local communities, bureaucrats, consultants …), as has been well documented in environmental politics (Forsysth 2003) including in water governance in Southeast Asia (Contreras 2007). The perceived validity of evidence also relates to recognitional justice and thus deeper processes of cultural, identity-based, and institutional bias.

Thus, EJ, and the particularities of how it is defined, can be understood as a framed narrative. Walker (2012: 40) seeks to systematise EJ research, acknowledging and accommodating the value of the plurality of the concept itself, and to add a critical edge. Walker is careful to differentiate between: evidence of how things are, which tends towards description; process of why things are how they are, which aims to be explanatory; and justice or how things ought to be, which is normative in prescription. Walker (2012: 5–8) also draws on the concepts of framing and claim making, drawing on concepts of methodological constructivism.

Furthermore, in the context of the nexus, the materiality of food, water, and energy, together with context, scale, and history shape how justice may be understood in any particular context. For example, regarding water Neal, Luka-siewicz, and Symec (2014: 1) state four characteristics with implications for interpreting the meaning of social and environmental justice:

- the spatial and temporal uneven distribution of water;
- the fact that water is essential for all life, with minimums needed for the survival of both the environment and humankind;
- water's added benefits to human well-being through the goods and services it provides;
- the ensuing political dimensions of power asymmetries affecting water governance.

Changes in water allocation (including due to infrastructure development), institutional arrangements, or hydrological regime (including due to climate change) can all raise issues of justice (Neal, Lukasiewicz, & Symec 2014). Groenfeldt (2010) recognises that all water policies are founded on ethical assumptions about water, yet there is a need to render these ethical values more explicit and imbue them with more of a sustainability perspective (including a recognition of climate change). Neal, Lukasiewicz, and Symec (2014: 1, 5–10), synthesising papers from a special issue journal titled "Why justice matters in water governance", propose four characteristics of water that hold implications for environmental and social justice:

- The physical properties of water: Water's distribution is uneven both spatially and temporally, therefore some places – at least before the intervention of institutions or technology – will hold a form of natural advantage in terms of access to or control over water resources
- Water is essential for all life: Both humans and the environment have a minimum (or basic) need for water for their survival. For humans, the Right to Water addresses the ability of humans to meet this basic need, whilst the concept of environmental flows is emerging as an approach to define and operationalise meeting the needs of nature.
- Water's benefits to human well-being: Beyond meeting basic needs, water provides added benefits to human well-being through the goods and services it provides. These include material benefits, health benefits, and human security benefits. There are a range of theories of justice (prior right theory, intergenerational justice …) that could guide how distribution of water might be "justly" prioritised beyond meeting basic needs, although which principle applies is often contested (Neal, Lukasiewicz, & Symec 2014: 8).
- Power asymmetries and water governance: Decision-making over access to and control over water is inherently political, and entail interaction between actors of asymmetrical power relations. Power shapes who defines – or who is heard to define – what is just as dominant discourses and worldviews are privileged over marginalised ones.

In EJ literature, scale, together with place and distance are important considerations. It can provide insight in to politicised nexused relationships not just between food, water, and energy systems but also across scales and considering the interests and agendas of actors. Political and economic geographers, for example, have highlighted how place-based specifics interact with higher scale actors, drivers, and structures, for example international markets and the investments and commodities that flow through them as well as national government policies, producing place-specific forms of environmental injustices (e.g. Harvey 1996; Leichenko & Solecki 2008; Sikor & Newell 2014). Local and national institutions mediate these forces and shape environmental (in)justice on the ground. The emergence of transnational EJ incorporated a concern for consumption patterns in Northern countries, international trade, and the deindustrialisation of the North and the relocation of polluting industries to the South (Schroeder, St. Martin et al 2008), as well as more recently processes of "land grabbing" and other natural resource appropriations (Borras & Franco 2012b). All of these higher-level processes have implications for the nexus, or indeed are nexused relationships themselves.

Regarding distance, EJ analysis also points towards the growing distance between the point of consumption and the point of production creating spatial disconnects that can render environmental injustices rendered in production areas invisible to the consumers (Agyeman 2014). General levels of inequity in consumption are demonstrated by the vastly different ecological footprints of countries of the North and the South, as well as between different groups within both.[5] Thus, a growing ecological footprint, together with a growing distance between consumption and production, significantly raises the likelihood of environmental injustice, including across nexused natural resources. Relatedly, other concepts that have emerged from EJ literature include NIMBYism (not in my back yard) and "the path of least resistance" (Bullard 1990: 4), whereby environmentally damaging projects are pushed away from areas with high social, economic, and political capital and towards areas with weaker capitals (Schelly & Stretesky 2009; Taylor 2000).

A final concept relevant, which has implications for the justice as distribution, procedure, and recognition, is the politics of scale (Norman, Bakker, & Cook 2012). This refers to how projects or issues are framed by multiple scales to (de)legitimise actors claims to environmental benefits and harms. Thus, a proposed large hydropower project, for example, may simultaneously be claimed as having net harm by communities who experience impact directly to their livelihoods at the local scale, whilst having net benefit by the State and/or its private sector developers as it seeks to ensure "national" energy security and contribute towards economic growth. Thus, in decision-making processes, many local scale impacts are rendered invisible (i.e. non-recognised) by project proponents (Dore & Lebel 2010).

In some ways similar to the nexus, EJ is a field of research and a policy arena, but – in contrast to the nexus' donor-led approach – is also a grassroots and transnational social movement. Academics, social activists, and even governments have conceived of, framed, and theorised EJ from plural normative and analytical perspectives. A shared concern, however, is an emphasis on social difference and how groups of people, differentiated for example by race, gender, or class, experience the environment differently (Robbins 2012:74).

EJ is at its strongest in evaluating fairness in decision-making, and explaining why (in)justices may have occurred. It is institutionally-rooted, with an emphasis on understanding processes of decision making and with strong linkages to policies, law, and systems of justice – a weakness of the current nexus approaches. EJ approaches, on the other hand, are arguably weaker in explaining inter-sectoral linkages between food, water, and energy systems, including consequences of cross-sectoral decisions that could have justice implications. We thus argue that in light of food, water, and energy trade-offs, bridging the gap between the nexus and EJ can redress in part a weakness of each.

6.3. Towards a rights-based approach to the nexus

The discourse on rights is based on the notions of rights holders and duty bearers. In the case of environmental resources this is complicated by notions of place and space, ownership and, in the case of water in particular, the fugitive nature of the resource across space and between communities. In many ways a combined approach to addressing water needs – the need to access quantities of sufficient quality to guarantee life and well-being – requires the use both of tools of EJ and a rights-based approach.

Water, food, and energy, are fundamental to meeting human needs. They relate to a range of positive human rights, including the right to health, food, and well-being. Thus, there is also the potential for a right-based approach to the nexus. However, there is a perception amongst some EJ advocates who say that rights-based approaches are too apolitical and local, ignoring the political process and how they are essentially dominated by global actors (Schlosberg 2004).

There has been growing momentum around the right to a safe, clean and healthy environment. Early on, the UN Draft Principles on Human Rights and the Environment began with these three statements:

1. Human rights, an ecologically sound environment, sustainable development and peace are interdependent and indivisible.
2. All persons have the right to a secure, healthy and ecologically sound environment. This right and other human rights, including civil, cultural, economic, political and social rights, are universal, interdependent and indivisible.
3. All persons shall be free from any form of discrimination in regard to actions and decisions that affect the environment.[6]

More recently, in his statement to the Human Rights Council in March 2014, the current UN Special Rapporteur on Human Rights and the Environment, Professor John Knox, concluded that: "I believe that it is now beyond argument that human rights law includes obligations relating to the environment" (Knox 2014). He had previously identified three types of environmental human rights obligations of States that broadly align with the principles of environmental justice (Knox 2012):

● Procedural obligations, including to assess environmental impacts, share information, facilitate public participation, and provide access to effective remedies for environmental harm.

- Substantive obligations to protect against environmental harm that interferes with the enjoyment of human rights, including adopting and implementing an appropriate legal framework that strikes a reasonable balance between environmental protection and other priorities. Substantive rights include life, health, food and water. It also includes the right to self-determination, for example when indigenous groups are threatened by hydropower projects.
- An obligation to take account of groups who may have particular vulnerabilities to environmental harm. This may include the impacts of environmental pollution to children's health, situations that may have disproportionate effects on women, and impacts on indigenous people who have a particularly close relationship with natural resources.

In recent years, various claims for justice have emerged related to individual components of the nexus. Food justice, also often linked to access to land and related natural resources, is advocated for within a social movement such as Via Campasina, as well as more institutionalised processes such as the UN's Special Rapporteur on the Right to Food. Various food justice concepts have emerged, for example Food Sovereignty and Land Sovereignty (Patel 2009; Agarwal 2014; Borras & Franco 2012a; Borras et al. 2011; Patel 2009). Regarding water, there have been equivalent movements, including against water grabbing, and in pursuit of the Right to Water (Franco et al 2014; Mehta, Veldwisch, & Franco 2012; Sultana & Loftus 2012). Meanwhile, more recently, questions have also been raised towards the production and distribution of energy, both in terms of the EJ impacts of the production of energy (Middleton 2012), and given the diversity of potential priorities for energy security (urban versus rural areas, household versus industrial buyers). Hildyard, Lohmann, & Sexton (2012) highlight that attaining national energy security is typically interpreted as energy to ensure economic growth, which is not necessarily equivalent to "energy for all" that prioritises energy to meet the whole population's basic needs. Thus "national energy security" is often used to justify energy enclosures, whilst masking energy inequalities, and creating new scarcities and insecurities as people are dispossessed of energy, food, water, land, and other necessities of life.

The view of a combined right to water and food including water for productive use for peasant farmer is increasingly discussed in international policy forums. A draft version of the General Comment 15 (GC15) of the Committee of Economic, Social and Cultural Rights in November 2002 had a section which stated that "the right to adequate food entitles an individual or group to secure the water necessary for the protection of food" (quoted in Winkler 2017: 121). Van Koppen et al asserts that GC15 "implies a right to water for livelihoods with core minimum service levels for water to homesteads that meet both domestic and small-scale productive uses" (Van Koppen et al 2017). This has been taken up by NGOs such as Bread for the World, and it is also reflected in the draft version of the UN Declaration on the Rights of Peasants and Other People Working in the Rural Area (United Nations 2018).

Given the transnational nature of some human rights and environment cases, and given also the uneven access to justice globally, the role of ETOs could be a significant transboundary rights-based justice arrangement and amendable to a nexus rights-based approach. ETOs are a recent evolution in international human rights law, and can be defined as "Obligations relating to the acts and omissions of a State, within or beyond its territory, that have effects on the enjoyment of human rights outside of that State's territory" (ETO Consortium 2013). ETOs are of particular importance when trans-national corporations operate in countries where "the accountability to human rights is low, the legal and institutional frameworks are too weak and much more favorable to (private) investors than victims/ local communities who are rarely consulted and heard" (ETO Consortium 2013). For example, the implications of ETOs for the Right to Water on transboundary rivers has only recently begun to be explored (Bulto 2011; Bulto 2014). In Southeast Asia, since the early 2010s, there has been growing interest in ETOs by civil society, as well as also be some national human rights institutions, as a means to challenge various intra-regional flows of investment in large-scale resource extraction projects that threaten or violate human rights (Middleton 2018).

Hoff (2011:5), in the background document to Bonn2011, considers that the nexus can facilitate a sustainability transition through: increasing resource use efficiency; generating knowledge that informs trade-offs and identifies synergies across sectors; investing to sustain ecosystem services; and accelerating access and integrating the poorest. The last dimension is linked to dimensions of justice and equity. Overall, any normative framing of the nexus – whether human rights-based or otherwise – must be embedded within the political economy of food, water, and energy; indeed any meaningful notion of justice must emerge from those within the water/food/energy system itself. Yet, the predominant analytical approach to "the nexus" at present has been more apolitical "socio-ecological systems thinking", although several researchers have recently sought to unpack the nexus through a more explicit political economy lens (Leck 2015).

6.4. Case study: Nexus, human rights and justice in South East Asia

In Southeast Asia, while individual government recognition of human rights is patchy and some governments appear to actively limit the diffusion of human rights-based approaches, there has also been some regional movement towards building a regional human rights system. At present there are National Human Rights Institutions (NHRIs) in Indonesia (founded 1999), Malaysia (founded 1999), Myanmar (founded 2011), the Philippines (founded 1987), and Thailand (founded 2001). The Paris Principles are an international benchmark by which NHRIs should work to protect and promote human rights, including with regard to their autonomy from the government. Indonesia, Malaysia, and the Philippines are presently rated as fully compliant (rating A) with the Paris Principles, Myanmar is partially compliant (rating B), and Thailand was downgraded from A to B in November 2015 (GANHRI 2017). However, even in the case of the A-rated NHRIs, the mandate of NHRIs in Southeast Asia are relatively weak; none, for example, are able to directly prosecute offenders in the court, but rather they carry out investigations and make recommendations (Grennan et al 2016).

At the regional scale, in October 2009 the governments established the ASEAN Intergovernmental Commission on Human Rights (AICHR), and subsequently, in November 2012, ASEAN adopted its Human Rights Declaration, in which Article 28 f states "Every person has the right to an adequate standard of living … including: … The right to a safe, clean and sustainable environment". Signalling its intent to pick up the Right to the Environment as a thematic focus, in September 2014 AICHR organised a two-day workshop titled "AICHR Workshop on Human Rights Environment and Climate Change", with follow-up workshops in September 2015 and October 2017. Meanwhile, AICHR has also published a thematic study titled "Baseline Thematic Study on CSR and Human Rights" that frequently refers to the environment and sustainability (Thomas & Chandra 2014). Civil society has also cautiously supported AICHR, whilst at the same time critiquing its lack of a protection role that means it is currently unable to receive and investigate cases (Gomez & Ramcharan 2014).

Regarding environmental governance, a recent study evaluated progress towards Principle 10 of the Rio Declaration in Southeast Asia for Thailand, Vietnam, Indonesia, and the Philippines (TEI 2011). Principle 10 addresses "access rights". which can also be understood as key principles of EJ. Overall, the study concluded that environmental governance is increasingly a recognised political agenda as compared to the past, and there is a broad trend towards increasing recognition of access to information, public participation, and access to justice including remedy and redress in the constitutions, although not necessarily with specific reference to the environment. Furthermore, increasingly comprehensive legislation regarding the environment is promulgated, alongside a tendency towards policies of decentralisation of political and administrative responsibility that open the possibility for better community management of natural resources. At the same time, the study identifies many wide gaps between the legal frameworks and implementation in practice, incoherent or incomplete legal frameworks, and limited capacities of the state and of civil society.

For example, the planning, construction, and operation of large hydropower dams hold implications for a wide range of substantive and procedural human rights (Hurwitz 2014), including human rights related to environment (Rieu-Clarke 2015a), and that epitomise the nexus relationships in play from the perspective of justice. In Southeast Asia, transnational civil society have increasingly utilised ETO cases as a means to investigate human rights threatened by hydropower projects on transboundary rivers involving cross-border investments, including via NHRIs (Middleton 2018; Middleton & Pritchard 2016). Recognising the challenges of seeking justice across borders, a range of arenas across scales have been utilised (Table 6.1), where each arena is understood as a politicised spaces of governance in which a process for claiming and/or defending rights or seeking redress for rights violations take place. Overall, Middleton and Pritchard (2016: 82) conclude that:

TABLE 6.1 Typology of legal "arenas of water justice" for human rights protection in Southeast Asia

Scale	Arena
National	National justice system National Human Rights Institution
Regional inter-governmental	ASEAN Intergovernmental Committee on Human Rights (AICHR) ASEAN Children and Women Commission
International inter-governmental	UN - Human Rights Council UN - Special Rapporteurs Universal Periodic Review Core treaties (Optional Protocol mechanisms – the Convention on the Elimination of all Forms of Discrimination Against Women (CEDAW), the Convention on the Rights of the Child (CRC) etc).
International voluntary/ non-binding mechanisms	Corporate policies of project developers / financiers Multi-lateral guidelines (e.g. Organisation for Economic Co-operation and Development (OECD) Standards on Multi-National Corporations) Multi-stakeholder voluntary processes (e.g. Hydropower Sustainability Assessment Protocol)

Source: Reproduced from Middleton & Pritchard 2016

> These arenas could be a key avenue offering access to justice when contestation emerges over shared resources within and across borders, which takes place in the context of uneven access to justice across the region. …. [However] These arenas are created, affirmed and reinforced … only through the innovative actions of affected communities, civil society groups, and allied individuals bravely challenging powerful actors often at personal risk.

Several regional conferences in Southeast Asia regarding transboundary water governance and the nexus have hinted at the need to address issues of access and inclusivity. In April 2014, for example, the Second Mekong River Commission Summit was held in Ho Chi Minh City, Vietnam (Bach 2014),[7] from which a nexus-framed message was delivered in person to the region's highest-level political leaders. It stated:

> In order to collectively benefit from the opportunities [of the nexus perspective], transboundary agreements and institutions develop and need to adapt to changing environments. For these to work effectively, a combination of political will, technical cooperation and an inclusive process is required.

Activists and scholars, however, have highlighted how these statements over transboundary waters in the past have not subsequently addressed issues of limited inclusiveness and accountability in practice (see Cooper 2012; Hansson, Hellberg,

& Öjendal 2012) and that greater attention must be paid to power asymmetries and deep politics rather than focus on technical solutions. Fundamentally, whilst many projects have promoted more inclusive water governance – often emphasising the need to hear marginalised communities' voice – they have not challenged too directly the region's underlying systemic injustices.

Many civil society and social movements in Southeast Asia rally around the rights of communities to access and manage their natural resources, including to rivers, forests, and land (Ahmed & Hirsch 2000; Cuasay & Vaddhanaphuti 2005; LRAN 2011). National and local civil society groups, however, have rarely explicitly utilised the nexus as a framing for their work to date. Despite this, implicit to many campaigns are nexus-type trade-offs, as has been demonstrated in heated debates around the revived plans for dams on the Mekong River's mainstream since 2007 (Grumbine, Dore, & Xu 2012; Matthews 2012; WWF 2012). It is in these movements that claims for justice are most commonly heard (Middleton 2012; Rieu-Clarke 2015).

Notes

1　See www.ejolt.org and www.ejatlas.org.
2　It includes access to information, participation in decision making, and access to justice systems.
3　A detailed treatment of justice is beyond the scope of this paper (see Walker, 2012: 42–53, and Schlosberg, 2007).
4　Walker (2012) argues this analytical move helps move beyond some past criticisms of EJ scholarship that it is overly descriptive, heavily framed by existing institutions and legislation, and, appearing to emphasise rule of law and access to justice rooted in a liberal political model with limited potential to accommodate legal plurality and radical transition.
5　However, the North is in the South, and the South in the North; namely that there are enclaves of high consumption in the South, whilst there are also areas of poverty and low consumption in the North.
6　http://hrlibrary.umn.edu/instree/1994-dec.htm (accessed 13 June 2018).
7　The International Conference on Cooperation for Water, Energy and Food Security Under Climate Change in the Mekong Basin, 2–3 April 2014, Ho Chi Minh City.

References

Agarwal, B. (2014). Food sovereignty, food security and democratic choice: critical contradictions, difficult conciliations. *The Journal of Peasant Studies*, 41(6), 1247–1268.

Agyeman, J. (2014). Global environmental justice or Le droit au monde? *Geoforum*, 54(0), 236–238.

Agyeman, J. (2005). *Sustainable communities and the challenge of environmental justice*. New York: NYU Press.

Agyeman, J. (2001). Ethnic minorities in Britain: short change, systematic indifference and sustainable development. *Journal of Environmental Policy & Planning*, 3(1), 15–30.

Agyeman, J., Bullard, R. D., & Evans, B. (2002). Exploring the nexus: Bringing together sustainability, environmental justice and equity. *Space and Polity*, 6, 77–90.

Agyeman, J., Bullard, R. D., & Evans, B. (2003). Introduction: Joined-up thinking bringing together sustainability, environmental justice, and equity. In J. Agyeman, R. D. Bullard, & B. Evans (eds). *Just sustainabilites: Development in an unequal world* (pp. 1–18). London: Earthscan.

Ahmed, M. & Hirsch, P. (2000). *Common property in the Mekong: Issues of sustainability and subsistence*. Penang, Malaysia, Sydney, Australia: ICLARM – The World Fish Center, Australian Mekong Resource Centre, Sida.

Bach, H., Glennie, P., Taylor, R., Clausen, T. J., Holzwarth, F., Jensen, K. M., Meija, A., & Schmeier, S. (2014). *Cooperation for water, energy and food security in transboundary basins under changing climate*. Lao, PDR: Mekong River Commission.

Baird, I. G., & Barney, K. (2017). The political ecology of cross-sectoral cumulative impacts: Modern landscapes, large hydropower dams and industrial tree plantations in Laos and Cambodia. *The Journal of Peasant Studies*, 44(4), 769–795.

Barney, K. (2007). Power, progress and impoverishment: Plantations, hydropower, ecological change and rural transformation in Hinboun District, Lao PDR. Toronto: York University Centre for Asian Research (YCAR) Working Paper No. 1.

Barney, K. (2009). Laos and the making of a "relational" resource frontier. *Geographical Journal*, 175(2), 146–159.

Batterbury, S. (2001). Landscapes of diversity: A local political ecology of livelihood diversification in south-western Niger. *Ecumene*, 8(4), 437–464.

Beder, S. (2006). *Environmental principles and policies: An interdisciplinary introduction*. London: Earthscan.

Benford, R. (2005). The half-life of the environmental justice frame: Innovation, diffusion and stagnation. In D. N. Pellow & R. J. Brulle (eds). *Power, justice and the environment: A critical appraisal of the environmental justice movement* (pp 37–54). Cambridge, MA: MIT Press.

Bernstein, H. (2010). *Class dynamics of agrarian change*. Halifax, Canada: Fernwood Publishing.

Boelens, R. & Zwarteveen, M. (2005). Prices and politics in Andean water reforms. *Development and Change*, 36, 735–758.

Boelens, R., Perreault, T., & Vos, J. (eds). (2018). *Water justice*. Cambridge, UK: Cambridge University Press.

Borras, S. M. Jr. & Franco, J. (2012a). A "land sovereignty" alternative? Towards a peoples' counter-enclosure. TNI Agrarian Justice Programme Discussion Paper. Amsterdam: Trans National Institute (TNI).

Borras, S. M. Jr. & Franco, J. (2012b) 'Global land grabbing and trajectories of agrarian change: A preliminary analysis', *Journal of Agrarian Change* 12(1): 34–59.

Borras, S. M. Jr., Hall, R., Scoones, I., White, B., & Wolford, W. (2011). Towards a better understanding of global land grabbing: An editorial introduction. *Journal of Peasant Studies*, 38(2), 209–216.

Bullard, R. (1990). *Dumping in Dixie: Race, class and environmental quality*. Bolder, CO, San Francisco, CA, and Oxford: Westview Press.

Bullard, R. D. (1994). *Unequal protection: Environmental justice and communities of color*. San Francisco, CA: Sierra Club Books.

Bulto, T. S. (2014). *The extraterritorial application of the Human Right to Water in Africa*. Cambridge, UK: Cambridge University Press.

Bulto, T. S. (2011). Towards rights-duties congruence: Extraterritorial application of the Human Right to Water in the African human rights system. *Netherlands Quarterly of Human Rights*, 29(4), 491–523.

Cifuentes, E., & Frumkin, H. (2007). Environmental injustice: case studies from the South. *Environmental Research Letters*, 2(4), 045034.

Clapp, J. & Dauvergne, P. (2011). *Paths to a green world: The political economy of the global environment*. Cambridge and London: MIT Press.

Contreras, A. P. (2007). Synthesis: Discourse, power and knowledge. In L. Lebel, J. Dore, R. Daniel, & Y. S. Koma (eds). *Democratizing water governance in the Mekong Region* (pp. 227–236). Chiang Mai: Mekong Press.

Cooper, R. (2012). The potential of MRC to pursue IWRM in the Mekong: Trade-offs and public participation. In J. Öjendal, S. Hansson, & S. Hellberg (eds). *Politics and development in a transboundary watershed: The case of the Lower Mekong Basin* (pp. 61–82). Dordrecht, Heidelberg, London, and New York: Springer.

Cuasay, P. & Vaddhanaphuti, C. (2005). *Commonplaces and comparisons: Remarking eco-political spaces in Southeast Asia*. Chiang Mai: Chiang Mai University.

Davis, M. (2014). Managing environmental systems: The water-energy-food nexus. Research Synthesis Brief. Stockholm: Stockholm Environment Institute (SEI).

Dore, J. & Lebel, L. (2010). Deliberation and scale in Mekong Region water governance. *Environmental Management*, 46(1), 60–80.

ETO CONSORTIUM (2013). *Maastricht principles on extraterritorial obligation of states in the area of economic, social and cultural rights*. Heidelberg: FIAN International.

Fisher, E. (2003). Sustainable development and environmental justice: Same planet, different worlds. *Environs*, 26, 201–217.

Foran, T. (2015). Node and regime: Interdisciplinary analysis of water-energy-food nexus in the Mekong region. *Water Alternatives*, 8(1), 655–674.

Forsysth, T. (2003). *Critical political ecology: The politics of environmental science*. London and New York: Routledge.

Franco, J., Feodoroff, T., Kay, S., Kishimoto, S., & Pracucci, G. (2014). *The global water grab: A primer*. Amsterdam: Trans National Institute (TNI).

GANHRI (2017). *Chart of the Status of National Institutions: Accreditation status as of 26 May 2017*. Geneva: GANHRI (Global Alliance of National Human Rights Institutions).

Gober, P., K.L. Larson, R. Quay, C. Polsky, H. Chang, & V. Shandas. (2013). Why land planners and water managers don't talk to one another and why they should! *Society and Natural Resources*, 26(3), 356–364.

Gomez, J., & Ramcharan, R. (2015). Introduction: Democracy and Human Rights in Southeast Asia. *Journal of Current Southeast Asian Affairs*, 33(3), 3–17.

Greacen, C. S. & C. Greacen (2012). *Proposed power development plan (PDP) 2012 and a framework for improving accountability and performance of power sector planning*. Bangkok: Palang Thai.

Grennan, K., Velarde, W. D. G., Zambrano, A. T., & Moallem, S. (2016). *The role of NHRIS to redress transboundary human rights violations: A Southeast Asian case study*. Raoul Wallenberg Institute of Human Rights and Humanitarian Law and Columbia University.

Groenfeldt, D. (2010). Viewpoint – The next nexus? Environmental ethics, water policies, and climate change. *Water Alternatives*, 3(3), 575–586.

Grumbine, E., Dore, J., & Xu, J. (2012). Mekong hydropower: Drivers of change and governance challenges. *Frontiers in Ecology and the Environment*, 10(2), 91–98.

Hall, D., Hirsch, P., & Li, T. M. (2011). *Powers of exclusion: Land dilemmas in Southeast Asia*. Singapore: NUS Press.

Hansson, S., Hellberg, S., & Öjendal, J. (2012). Politics and development in a transboundary watershed: The case of the Lower Mekong Basin. In J. Öjendal, S. Hansson, & S. Hellberg (eds). *Politics and development in a transboundary watershed: The case of the Lower Mekong Basin* (pp. 1–18). Dordrecht, Heidelberg, London, New York: Springer.

Harvey, D. (1996). *Justice, nature and the geography of difference*. Oxford: Blackwell Publishers.

Hildyard, N., Lohmann, L. & Sexton, S. (2012). *Energy security: For whom? For what?* Sturminster Newton, UK: The Corner House.

Hoff, H. (2011). *Understanding the nexus*. Background Paper for the Bonn 2011 Conference: The Water, Energy and Food Security Nexus. Stockholm: Stockholm Environment Institute.

Hurwitz, Z. (2014). *Dam standards: A rights-based approach - A guide book for civil society*. Berkeley, CA: International Rivers.

Knox, J. H. (2014). Report of the Independent Expert on the issue of human rights obligations relating to the enjoyment of a safe, clean, healthy and sustainable environment. John H. Knox: Mapping report. Human Rights Council: Twenty-fifth session, Agenda item 3, Promotion and protection of all human rights, civil, political, economic, social and cultural rights, including the right to development. New York: United Nations Human Rights Council.

Knox, J. H. (2012). Report of the Independent Expert on the issue of human rights obligations relating to the enjoyment of a safe, clean, healthy and sustainable environment. Human Rights Council: Twenty-second session, Agenda item 3, Promotion and protection of all human rights, civil, political, economic, social and cultural rights, including the right to development. New York: United Nations Human Rights Council.

Lahiri-Dutt, K. (2014). Beyond the water-land binary in geography: Water/lands of Bengal re-visioning hybridity. *ACME*, 13(3), 505–529.

Lavau, S. (2013). Going with the flow: Sustainable water management as ontological cleaving. *Environment and Planning D: Society and Space*, 31, 416–433.

Leach, M., Stirling, A. C., & Scoones, I. (2010). *Dynamic sustainabilities: Technology, environment, social justice*. London: Routledge.

Leck, H., Conway, D., Bradshaw, M., & Rees, J. (2015). Tracing the water–energy–food Nexus: description, theory and practice. *Geography Compass*, 9(8), 445–460.

Leichenko, R. M. & Solecki, W.D. (2008). Consumption, inequity, and environmental justice: The making of new metropolitan landscapes in developing countries. *Society & Natural Resources*, 21(7), 611–624.

Lele, U., Klousia-Marquis, M., & Goswami, S. (2013). Good governance for food, water and energy security. *Aquatic Procedia*, 1(0), 44–63.

LRAN (Land Research Action Network) (2011). *Defending the commons, territories and the right to food and water*. Quezon City, the Philippines: Focus on the Global South.

Martinez-Alier, J. (2014). The environmentalism of the poor. *Geoforum*, 54, 239–241.

Matthews, N. (2012). Water grabbing in the Mekong basin: An analysis of the winners and losers of Thailand's hydropower development in Lao PDR'. *Water Alternatives*, 5(2), 392–411.

Mehta, L., Allouche, J., Nicol, A., & Walnycki, A. (2014). Global environmental justice and the right to water: The case of peri-urban Cochabamba and Delhi. *Geoforum*, 54, 158–166.

Mehta, L., Veldwisch, G. J., & Franco, J. (2012). Introduction to the special issue: water grabbing? focus on the (re)appropriation of finite water resources. *Water Alternatives*, 5(2), 193–207.

Middleton, C. (2012). The "nature" of beneficial flooding of the Mekong River. *Social Science Journal*, 42(2), 180–208.

Middleton, C. (2012). Transborder environmental justice in regional energy trade in Mainland South-East Asia. *Austrian Journal of Southeast Asian Studies*, 5(2), 292–315.

Middleton, C.; Allouche, J.; Gyawali, D., & Allen, S. (2015). Where's the (environmental) justice? The rise and implications of the water-energy-food nexus in Southeast Asia. *Water Alternatives*, 8(1), 627–654. Downloadable at: www.water-alternatives.org/index.php/alldoc/articles/vol8/v8issue1/269-a8-1-2/file.

Middleton, C. & Pritchard, A. (2016). *Arenas of water justice on transboundary rivers: A case study of the Xayaburi Dam, Laos. Water governance dynamics in the Mekong Region*. Petaling Jaya: Strategic Information & Research Development Centre.

Middleton, C. (2018). National human rights institutions, extraterritorial obligations and hydropower in Southeast Asia: Implications of the region's authoritarian turn. *Austrian Journal of Southeast Asia Studies*, 11(1), 81–97.

Mutz, K., Gary, C. B., & Douglas, S. K. (eds) (2002). *Justice and natural resources: Concepts, strategies, and applications*. Washington, DC: Island Press.

Neal, M. J., Lukasiewicz, A., & Symec, G. J. (2014). Why justice matters in water governance: Some ideas for a "water justice framework". *Water Policy*, 16, 1–18.

Norman, E. S., Bakker, K., & Cook, C. (2012). Water governance and the politics of scale. *Water Alternatives*, 5(1), 52–61.

Nuntavorakarn, S. & D. Sukkumnoed, D. (2008). *Public participation in renewable energy development in Thailand: HIA Public scoping and public review of the two controversial biomass power plant projects. HIA Development in Thai Society "Empowering People Ensuring Health.* The Asia and Pacific Regional Conference on Health Impact Assessment, 8–10 December 2008. Chiang Mai, Thailand. HPPF. Bangkok, Healthy Public Policy Foundation (HPPF).

Okereke, C. (2008). *Global justice and neoliberal environmental governance: Sustainable development, ethics and international co-operation*. London and New York: Routledge.

Patel, R. (2009). Food sovereignty. *The Journal of Peasant Studies*, 36(3), 663–706.

Rieu-Clarke, A. (2015). Transboundary Hydropower projects seen through the lens of three international legal regimes: Foreign investment, environmental protection and human rights. *International Journal of Water Governance*, 3(1), 27–48.

Robbins, P. (2012). *Political ecology*. Chichester, UK: Wiley-Blackwell.

Roberts, A. S. (2015). Lost in transition: Landscape, ecological gradients, and legibility on the Tonle Sap floodplain. In S. Milne & S. Mahanty (eds). *Conservation and development in Cambodia: Exploring frontiers of change in nature, state and society* (pp. 53–74). London and New York: Routledge.

Roth, D., Boelens, R., & Zwarteveen, M. (2015). Property, legal pluralism, and water rights: The critical analysis of water governance and the politics of recognizing "local" rights. *The Journal of Legal Pluralism and Unofficial Law*, 47(3), 456–475.

Schelly, D. & Stretesky, P. B. (2009). Insights and applications: An Analysis of the "path of least resistance" argument in three environmental justice success cases. *Society and Natural Resources*, 22, 369–380.

Schlosberg, D. (2004). Reconceiving environmental justice: Global movements and political theories. *Environmental Politics*, 13(3), 517–540.

Schroeder, R., St. Martin, K., Wilson, B., & Sen, D. (2008). Third world environmental justice. *Society and Natural Resources*, 21, 547–555.

Sikor, T. & Newell, P. (2014). Globalizing environmental justice? *Geoforum*, 54(0), 151–157.

Simpson, A. (2007). The environment-energy security nexus: Critical analysis of an energy "love triangle2 in Southeast Asia. *Third World Quarterly*, 28(3), 539–554.

Smajgl, A. & Ward, J. (2013). A framework for bridging science and decision making. *Futures*, 52(8), 52–58.

Sneddon, C. & Fox, C. (2008). Struggles over dams as struggles for justice: The World Commission on Dams (WCD) and Anti-Dam Campaigns in Thailand and Mozambique. *Society & Natural Resources*, 21(7), 625–640.

Sultana, F. & Loftus, A. (eds) (2012). *The right to water: Politics, governance and social struggles*. London: Sterling and VA: Earthscan.

Taylor, D. E. (2000). The rise of the environmental justice paradigm. *Am Behav. Sci.*, 43(4), 508–580.

TEI (2011). Environmental governance in Asia: Independent asessments of national implementation of Rio Declaration's Principle 10. Bangkok: Thailand Environment Institute (TEI).

Thomas, T. & Chandra, A. (2014). *Thematic study on CSR and Human Rights*. Jakarta: ASEAN Intergovernmental Commission on Human Rights (AICHR).

United Nations (2018). A/HRC/WG.15/5/2. Revised draft of the United Nations declaration on the rights to peasants and other people working in rural areas. Human Rights Council.

Van Koppen, B., Hellum, A., Mehta, L., Derman, B., & Schreiner, B. (2017). Rights-based freshwater governance for the twenty-first century: Beyond an exclusionary focus on domestic water uses. In E. Karar (ed). *Freshwater governance for the 21st Century* (pp. 129–143). Amsterdam: Springer.

Walker, G. (2012). *Environmental justice: Concepts, evidence and politics.* London and New York: Routledge.

Weitz, N., Nilsson, M., Huber-Lee, A., & Hoff, H. (2014). *Cross-sectoral integration in the Sustainable Development Goals: A nexus approach.* SEI Discussion Brief. Stockholm: Stockholm Environment Institute.

Winkler, I. T. (2017). Water for food: a human right perspective. In M. Langford and A. F. S. Russell (eds). *The human right to water: Theory, practice and prospects* (pp. 119–143). Cambridge, UK: Cambridge University Press.

WWF (Worldwide Fund for Nature) (2012). *Mekong dams could rob millions of their primary protein source.* Available at: wwf.panda.org/?206033/Mekong-dams-could-rob-millions-of-their-primary-protein-source (accessed on 20 January 2015).

Yoo, Y. (2013). *Renewable energy development and environmental justice in Thailand: Case studies of biomass energy projects in Roi-Et and Suphanburi Provinces.* Master's Thesis. Bangkok: Chulalongkorn University.

7

ETHICS AND THE NEXUS

7.1. Nexus encounters ethics

Humans cannot survive without food, water, and energy, with water and food in particular understood as basic needs and a human right (Knox 2014). Whilst they may be afforded an economic value, they are also essential to life and thus their value transcends economic value alone. Indeed, when examining different water-related conflicts such as with many dam projects around the world, protagonists approach the issue from very different value premises with little common ground between them. Hydrocracies promoting the dam argue for benefits from the sale of electricity, while activists protesting the project see it as destroying natural habitats as well as sacred sites of indigenous communities upon which no monetary value can be placed or trade-off considered. It is precisely this additional complexity that arises as diverse concerns interact, and which is what the nexus approach is all about, that ethical concerns come into play and cannot be ignored. Indeed, Hilary Putnam (2005) describes ethics as "concerned with the solution of practical problems, guided by mutually supporting but not fully reconcilable concerns". He further argues that these concerns come with real-life tensions that will not yield to a simple ethics "as a noble statue standing atop a single pillar". He prefers, much like a good nexus approach protagonist though without being a part of the nexus debate, to see it as a table with many legs, which wobbles a lot on an uneven surface, but is very hard to turn over.

Values[1] are at the centre of individual decisions, collective actions, public policies, and broader governance systems, and ethics address how humans ought to think and act, especially when there are competing values in play (Groenfeldt & Schmidt 2013). Deeply held values, both individual and collective, are what guide our actions, and in thus acting, we come face-to-face with the righteousness (or lack thereof) of the course we take. A quote often mentioned is Goethe's: "Knowing is not enough; we must apply. Willing is not enough; we must do." It is in the interplay of this knowing, willing, and acting that ethical conundrums make their presence felt, more so within a complex nexus context where different knowings give rise to dissimilar actions, often in conflict with each other. This conundrum, and the particular orientation of nexus governance, is not apparent without an explicit consideration of values and their associated ethics (Molle 2008; Pahl-Wostl, Gupta, & Petry 2008).

Rendering contesting values visible – including across scales of governance, between universal and situational scopes, across the structures of food, water and energy systems, and within the diversity of nature–society relations – does not in and of itself resolve ethical dilemmas, but it does focus our attention on to why ethics matter. Taking water as an example, Groenfeldt and Schmidt (2013: 14) point out: "Nobody denies the value of water, yet explicit attention to ethics has largely been absent in discussions of water governance." Extending Schmidt's (2010: 4) analysis of disagreements that might arise over water to a nexus frame, potential dilemmas or disagreements include:

1. Claims about facts or states of affairs, such as those about adequate water, food, and energy quantity or quality, including how "adequate" is defined;
2. Claims about distributional fairness in terms of benefits, costs, and risks, including the way that trade-offs are addressed between nexused sectors;
3. Claims about correctly ordered social relationships, such as whether water, food, and energy should be allocated according to economics or on the basis of factors such as human rights or rights to property or healthy ecosystems;
4. Claims about personal experiences, such as water's and food's significance to people of a particular culture or belief.

In the mainstream World Economic Forum (WEF) nexus literature to date, ethics have been little if at all been discussed explicitly (see Schlör et al 2018 for an exception). However, related work has explored aspects of nexus, for example, between water and climate change (Groenfeldt 2010), and water and food security (Lopez-Gunn, De Stefano, & Llamas 2012). Nexus approaches cannot avoid value judgements, and therefore implicitly neither can they avoid engaging in ethics, even when those involved do not acknowledge that they are doing so.

7.2. Values and ethics

Ethics address how humans ought to reflect and act, especially when there are competing values at play giving rise to conundrums and dilemmas when individuals and societies have to take decisions. What was previously called moral philosophy and, now ethics, asked, as Nowell-Smith (1954) puts it: "What shall I do?" and "What moral principles should I adopt?" In this sense, ethics are both a practical and a theoretical science: it mainly tries to answer practical questions and then to appraise and evaluate behaviour and customs according to its reasoning. But this distinction gives rise to the philosophical predicament first highlighted by David

Hume, who showed that reason and empiricism, while important, only went part of the way to describe the ends. Reason alone was impotent to produce any action towards an end which could be produced only by passion and feelings that drives humans to act, which again is the realm of morals and ethics (Malik 2014). The eminent philosopher G.E. Moore (1959) puts the essence of ethical questions into two types that will be answered very differently:

> whether something ought to exist for its own sake, is good in itself or has intrinsic value; and exactly what it is that we ask about an action, when we ask whether we ought to do it, whether it is a right action or a duty.

Ethics mostly tried to answer the second of these, i.e. what is the right action we should engage in, although with the rise of the environmental movement since the 1970s, the first question – of even things that have no immediate economic value having a right to exist and indeed even flourish under normal conditions – have come to occupy a more salient place in ethical debates. As Warnock (1998) puts it:

> The ethical, then, arises when someone begins to see that he must postpone his immediate wishes for the sake of the good. And "the good" here embraces both his own goodness, and the goodness of society of which he is a member.

These are old questions that all the major religions have grappled with through the ages, and which, despite some dalliance with modernity during much of the 20th Century, have come back to haunt us as complex social and environmental issues, such as persistent poverty, hunger, pollution, or climate change – the very stuff of the nexus approach – demand answers and actions. Looking at the Judeo-Christian tradition from as far back as Thales of Miletus, Sison (2009) examines the Biblical injunction of man created in the image of God and given privileged command over the rest of the creatures found in nature. While on the one hand this has given rise to the controlling attitude on the part of the humans, especially with more advanced and powerful technologies, the pre-Socratic origins of Judeo-Christianity also has, on the other hand, a powerful contemplative attitude that requires wonder and respect for nature. This second attitude is being revived in the post-modern and Green critiques of modernity that question the reification of technological progress.

Examining the case of Islamic water ethics, Hefny (2009) describes the key tenets for this major global religion which originated in water-scarce desert as based on stewardship and on the notion of the public good. Although humans are the most favoured among God's creatures, it is incumbent upon them to ensure that God's gift to humanity (i.e. nature) is well conserved and made equitably available for all on planet Earth, including future generations. Humans have no right to upset the natural order through pollution or over-exploitation of natural resources. He describes how, because of the utmost importance attached to personal cleanliness and public hygiene, a special council of leading Islamic scholars had to be called to issue an injunction (*fatwa*) that treated wastewater can theoretically be used for all purposes. The idea of water pricing is a vexing issue in Islam because it is seen as primarily a public good which should neither be bought nor sold. However, given the impracticality of such a rigid interpretation, Islamic scholars have divided a natural resource such as water into public property in its natural state with free access to all; restricted private property where owners have certain rights but also obligations not to pollute or hold back surplus; and private property developed through investments. In these divisions, the most important aspect is that water use was prioritised with quenching thirst having the highest precedence followed by domestic use and then irrigation.

In Hindu-Buddhist cosmology,[2] extreme monism and pantheism in normal social life jostle with each other; and ethics are defined by the term *dharma* (the correct way) which is dependent upon clan, caste, profession, stage of an individual's life, and the level of spiritual attainment. In this panoply, what is all right for one would be wrong for another. This gives rise to a pluralism of ethical relativity that is vexing to those more comfortable in monotheistic traditions. Examples given are that it is right for a king to kill someone (if the person has committed a heinous crime according to the law) but wrong for someone who does not carry the duty of a ruler; celibacy is obligatory for students but wrong for householders and so on, with higher ethical standards for those with more responsibilities (Gyawali 2009). The benchmark of Hindu ethics is encapsulated in the *Bhagwat Gita*, a section of the epic *Mahabharata* which is played out against the backdrop of a major battle. It is all about ethical conundrums faced by protagonists in the different warring parties as they engage in actions that call into question different moral injunctions. The essential point is that the overall cosmic order has to be preserved by observing *deshachara* (conduct befitting a locality), *lokachara* (conduct befitting a community), and *kulachara* (conduct befitting a family or a particular clan). These ethical complexities were possible to be put in practice in an isolated Hindu country that was predominantly agrarian but has been stressful when Hindu society has come under pressures of modernity with industrial urbanisation. Colopy (2012) explains in bewildering detail how Hindu society that considers rivers as goddesses and sacred seems to have no problem making them some of the most polluted in the world.

The important point to keep in mind as we conduct this quick overview of what ethics mean, in general and according to some of the major world religions, to nexus is that an understanding of values will be critical to implementing the nexus approach. And values are socially and historically determined in different lands, communities, and times. A nexus approach will in all probability provide better insights about the complex interlinkages between different sectors; but providing prescriptive solutions will have to be finely tailored to particular contexts.

7.3. Water, ethics, and IWRM

Focusing on water, in 1997 the United Nations Educational, Scientific, and Cultural Organization's (UNESCO) World Commission on the Ethics of Scientific Knowledge and Technology (COMEST) commenced a study on the ethics of freshwater (Selborne 2000). As summarised by Schmidt (2010), the report addressed three themes: (1) a sense of shared purpose and harmony with nature, (2) a balance between traditional human values and technological innovation, and (3) harmony between "the sacred and utilitarian in water, between the rational and the emotional." More studies followed; for example, a book titled *Water Ethics* emerged from the Marcelino Botín Water Forum in 2007 (Llamas, Martinez-Cortina, & Mukherji 2009). A series of essays addressed topics ranging from cultural traditional approaches on water ethics, to water as a human right and as an economic resource, and the ethics of water management, including that of integrated water resources management (IWRM), water rights, and water governance.

Whilst an ethics of water, food, and energy in a contemporary sense could be associated with the managerialism of administrative regulation and principles of distributional justice, a diversity of ethics of water and food is also embedded in many traditions, religious practices, informal institutions, and other social practices (Schmidt 2010).

IWRM, which emerged as the dominant global paradigm for water governance in the 1990s, holds a viewpoint on water management that ascribes water with a particular set of values (human right; economic) that in turn reflects a particular ethic (Molle 2008). Its core values have remained relatively constant as it has diffused globally by a range of actors including academics, donors, INGOs, and local officials, even as it has been adapted to a degree to particular country contexts (Allouche 2016). The four core principles of IWRM are detailed in the 1992 "Dublin Statement on Water and Sustainable Development" as:

- Fresh water is a finite and vulnerable resource, essential to sustain life, development and the environment
- Water development and management should be based on a participatory approach, involving users, planners and policy-makers at all levels
- Women play a central part in the provision, management and safeguarding of water
- Water has an economic value in all its competing uses and should be recognized as an economic good

Whilst framed by water professionals as a neutral managerial approach to water management and its integration, IWRM is of course highly positioned (Schmidt 2010). The statements above include subjective framings, such as that of "development", as well as contested notions such as water holding an economic value (versus non-comparable values, such as the sacred or social). Meanwhile, some important values are left unstated, such as who should be considered legitimate within participatory processes. Furthermore, whilst participation is claimed as a fundamental value of the IWRM approach, the role of the expert is also privileged thus giving a particular value to scientific forms of knowledge. As Schmidt (2010: 9) astutely argues: "... though integration may appear a neutral term, there are in fact numerous ethical judgments within IWRM."

7.4. Energy ethics

Energy justice, and an associated energy ethics, has been an emerging field for at least the past two decades, first by civil society groups since the 1990s and, later, within academic literature since at least 2010 (Heffron & McCauley 2017). Some have sought to draw upon the environmental justice framework as applied in particular to energy systems, namely distributional justice, procedural justice, and recognitional justice (Jenkins et al 2016; see Chapter 6). Meanwhile, Sovacool and Dwokin (2015) propose eight factors for decision-making in an energy justice framework: availability; affordability; due process; good governance; sustainability; intergenerational equity; intragenerational equity; and responsibility. Broadly speaking, this body of work has sought to engage at the energy system level, and on energy policy, to ensure that all individuals, across all areas, with safe, affordable, sustainable, and secure energy sources whilst addressing the unequal distribution of harms (Heffron & McCauley 2014), and how socially just energy transition might occur (Jenkins, Sovacool, & McCauley 2018). Given how the distance between the harms of energy production and the benefits consumption has often been suggested as one explanatory factor in the perpetuation of environmental injustice, work on energy justice has also emphasised the need to consider justice at each step of the supply chain from production through to consumption (Heffron & McCauley 2014).

Frigo (2017) shows how a traditional energy paradigm that emerged since the 18th Century, in which energy is understood as the "capacity to do work", is associated with a reductionist and mechanistic natural sciences in the language of mathematics and engineering. Thus, it has lent itself towards an instrumental and anthropocentric ethics. Frigo (2017: 7) argues that this traditional energy paradigm and its associated energy ethics "… has been propagated throughout the world via cultural, socio-economic, and techno-scientific colonization." It has limited the possibility to consider alternative worldviews and ethical approaches to energy as a multi-faceted phenomenon. To explore the possibilities for moving beyond the traditional energy paradigm Frigo (2017: 9) differentiates between "energy ethics" and the "ethics of energy" where: "ethics of energy appear intrinsically connected to a normative attitude, prescribing what we should do; energy ethics, conversely, move from and rest on descriptive grounds without suggesting any specific moral conduct." The former considers ethics as normative and plural within particular situational contexts, and can provide guidelines for resolving policy dilemmas through moral guidance, whilst the latter emphasises how different actors think about and experience energy revealing the complexity of contemporary energy debates.

As we discuss below, rather than deferring to universalist ethics paradigms, there is now a growing appreciation of the need for a plural situated ethics in energy debates, that should be taken seriously if persistent social tensions over meeting energy needs are to be resolved fairly (Mason & Mibourne 2014). This has included a significant broadening of the field, as we discuss below, to engage non-

Western theorists (Sovacool et al 2017), an emphasis on the lived complexities of everyday practices of energy (Ho 2015), and an extension of ethics from the human to incorporate the more-than-human world (Pinker 2018).

7.5. Food ethics

The topic of food ethics has sparked a lot of debates and controversies. Significant changes in the system of food production, such as the introduction of pesticides, artificial fertilisers, preservatives, genetic modification, and other forms of biotechnology, produced a whole range of ethical debates. As put by Comstock (2010), ethically justifiable conclusions inevitably rest on two kinds of claims: (a) empirical claims, or factual assertions about how the world *is* – claims ideally based on the best available scientific observations, principles, and theories, and (b) normative claims, or value-laden assertions about how the world *ought to be* – claims ideally based on the best available moral judgments, principles, and theories.

These debates about GMOs, and other technologies are reflective of broader structural ethical and philosophical questions with respect to our conception of time and space. As put by Korthals (2015), various technologies have enabled the continuous reduction of food processing time but these have created a distance between the producer and the consumer, while at the same time degraded the land and soil. The disconnect between food consumers and food production creates fundamental ethical dilemmas. Food comes from all over the world and there is little real-life experience among everyday (especially urban) consumers with modern farming. This has led not just to a physical distance but also a mental distance between the producer and the consumer (Brom 2000).

These gaps have critical consequences for the way consumers perceive products, and for the way they build their trust. Trust is no longer shaped by direct human interaction (Brom 2000). Furthermore, the evolutionary advantages to reduction of food collecting, producing, and digesting time, also produce the risk that people become alienated from food and become subordinated to corporate production of food. In the end, people no longer know what to buy and what they eat – but of course they must eat (Korthals 2015).

The other important ethical dilemma is the idea of abundance due to the easy digestibility of cooked food. The idea that we can eat without limit is because humans are constantly cheated by both their bodies and sciences. Senses do not signal the number of calories that are digested and the food sciences are, until recently, misleading in measuring uniform units of calories, independent of what form the food is eaten and digested. As a result, overweight and obesity are becoming serious public health concerns (Korthals 2015). These debates can be seen as very Western-centric, given the issue of food insecurity in the global south, although these are also becoming important public health issues in these countries.

7.6. Environmental ethics: From anthropocentric to non-anthropocentric ethics

Ethics matter as human actions towards each other and the more-than-human could have consequences for human and more-than-human[3] lives. In the field of environmental ethics, the relationship between humans and nature (or the "more-than-human") is of key concern.[4] It is asked, for example: Does nature have a value extending beyond its role to serve human needs?; Do some parts of nature have greater value than others?; What responsibilities do humans have towards nature? (Carter 2007). Clearly a normative framework is in existence, even as it is relatively less examined in comparison to studies of environmental governance systems and institutions. Whether these norms are claimed to be universal or situated is often also a key issue at stake.

There has been a growing attention to environmental ethics in general (Carter 2007; Light & Rolston 2002; Raj et al 2010) as well more specifically to water ethics (Brown & Schmidt 2010; Liu et al 2011), food ethics (Brom 2000; Comstock 2010; Korthals 2015; Zwart 2000), and energy ethics (Moss et al 2011). However, ethical questions with respect to all these sectors can be traced back to ancient civilization and religious beliefs (see Zwart 2000). Interestingly, these ethical concerns have always been discussed and debated from a sectoral perspective. The interrelationship between these sectors, in other words the nexus, has been far less considered from an ethical perspective.

Whilst governance paradigms continue to evolve for each sector – for example, adaptive management, resilience, and human security – inescapably, each approach must engage with ethics. As mapped out by Brown and Schmidt (2010), a range of ethics frameworks have been deployed. The general shift from an anthropocentric to ecocentric perspective is significant to the nexus from an ethical perspective as it invites a consideration of a wider range of relationships and how humans should value them.

Human dominion is used to legitimise claims to resources whereby humans – and in particular organised human conglomerations that have the power and resources to make those claims – take priority. Brown and Schmidt (2010: 19) note how these approaches are often critiqued as "anthropocentric, instrumental and patriarchal." Well known, for example, is the critique of Judeo-Christian ethics as anthropocentric "mastery over nature", whereby nature is treated as existing for instrumental purposes to meet human needs. It has also often been associated with patriarchal social ordering, as well as marginalising the moral standing of indigenous cultures. Furthermore, many of the literature on the topic is Western-centric.

Utilitarianism, namely the greatest good (pleasure or happiness over pain or unhappiness) for the greatest number, has a long tradition from the work of Jeremy Bentham and John Stuart Mills in the 19th Century. This ethic is also largely associated an anthropocentric viewpoint, whereby nature is rendered a resource that increase human happiness. Peter Singer (1975), however, has sought to extend the moral community counted within the utilitarian calculus, arguing that sentience is the pre-requisite for having "interests" and that the equal consideration of interests should be applied to all animals that can suffer, thus shifting beyond an anthropocentric ethic.

Summarising two dominant approaches to utilitarianism, Brown and Schmidt (2010: 80) write:

> Bureaucratic approaches to utility are characterized by their reliance on a centralized authority and experts who, for instance, engineer large dams or diversion projects in order to provide increased benefits to society. Alternately, individualist approaches are characterized by reliance on markets and the idea that there is, by definition, a gain in utility from all voluntary transactions.

Regarding the latter, it has been used as an ethical claim for the reduced role of the state and a growing role for the market, including the economic valuation of water, food, and energy. In nexus logic, there is evidence of both state-led and market-orientated approaches, and that thus draw on principles of anthropocentric utilitarianism as an ethical underpinning. As discussed further below, such universalist utilitarian approaches have been argued for their lack of attention to who is marginalised in the utilitarian calculus of the greatest good for the greatest number (Boelens, Vos, & Perreault 2018), otherwise known as the "tyranny of the majority". Related to this, the application of utilitarian principles can affect and displace pre-existing local governance arrangements based on systems of values existing situationally (Swyngedouw 2005). A second concern relates to determining whether the greatest good is calculated at the level of the aggregated individual, or at higher scales of various collective goods such as the protection of a watershed (Brown & Schmidt 2010). A third concern addresses distribution of resources, and if they should be distributed equitably as this also has ethical implications (Groenfeldt & Schmidt 2013). In the food sector for example, enough food is produced for the global population but more than one billion people on Earth suffer from hunger and even more from malnutrition. Furthermore, it has been estimated[5] that 15–50 per cent of food crops are lost between production and the market globally (and in developing countries as high as 65 per cent). This translates into the wastage of almost a quarter of all freshwater, crop area, and fertiliser currently used for food production! Moreover, while much of the food wasted in developed countries is at the household level, in developing countries it is at the level of field and transportation. This leads to a number of key ethical questions: Do well-fed people have an obligation to help the hungry and reduce their own waste? What should be the role and responsibilities of states and international bodies? Is more production necessary or should the existing food stocks be more justly distributed? What can be done about malnutrition and who has responsibility to tackle this problem? (Korthals 2015).

Another range of ethics emerges from the commons, defined by Elinor Ostrom (1994) as "… natural or human constructed systems that generate a finite flow of benefits in which: 1) exclusion of beneficiaries through physical and institutional means is especially costly; and 2) exploitation by one user reduces resource availability for others." Ethics emerge from the commons governance whereby its collective-action character entails various arrangements for multi-scaled and multi-actor decision-making, ranging from the local level to the global level, and that mix public and private institutions in the form of hybrid systems (Ostrom et al 1999).

Whilst Ostrom largely interprets the governance of the commons through game theory and thus draw on institutional economics and the rational interaction of individual actors, others highlight how the commons is a broader set of non-commoditised social relationships that incorporate a complex set of relationships, beliefs, norms, and interests of resource users (Bollier & Helfrich 2012). Surhardiman et al (2017: 11) argue that "Unpacking power and politics in water governance is crucial for understanding current challenges faced by the 'new commons' or common pool resources that require multi-scalar governance structures and mechanisms for their sustainable management." Under the rubric of "environmental pragmatism", other approaches that emphasise value pluralism, deliberative dialogue, and experimental environmental policies are also emerging that places community collaboration as central to the maintenance of local commons (Brown & Schmidt 2010).

At the transboundary level, the UN Water Courses Convention (UNWC) addresses principles for cooperation on transboundary rivers. It was adopted in 1997 and came into force in August 2014 after Vietnam became the 35th country to ratify it. Key principles include: on "equitable and reasonable utilization", the obligation "not to cause significant harm", to exchange data, and to cooperate. Overall, watercourse states enjoy equal rights to the utilisation of an international watercourse, drawing on the principle of "limited territorial sovereignty". The complexity of competing values multiply at the transboundary scale (Groenfeldt & Schmidt 2013). The same ethical questions can be asked for food and energy, where some have argued that the unsustainability of the current food and agricultural system puts the burden on the shoulders of taxpayers and nature as all the material and immaterial costs are externalised.

Another set of ethical perspectives broadens the scope of moral standing to incorporate the more-than-human where the intrinsic value of non-human life and non-life is acknowledged (e.g. Preiser, Pereira, & Biggs 2017; Schmidt 2017). Non-human of course includes animals, and the ethical dilemmas linked to intensive livestock farming, and where the animal is downgraded to a mere object, which has no voice, enduring pain and suffering by being confined in very small pens, and subject to inhumane forms of slaughtering. But more broadly, the scope of ethics in terms of non-life shifts from an anthropocentric one to an ecocentric one. Debates here relate not only to what has intrinsic value (and therefore moral standing within environmental ethics), but also what the implications are for human behaviour in relation to the more-than-human. One key writing has been Aldo Leopold's essay "The Land Ethic", in which he expanded the community to be considered ethically to include humans, non-human life, and also other components such as the soil and water (i.e. the biotic community). Leopold's land ethic emphasises the interconnections between these, where "A thing is right when it tends to preserve the integrity, stability, and beauty of the biotic community. It is wrong when it tends otherwise" (Leopold 1949). Armstrong (2009) has made an analogous argument in favour of a "water ethic". Brown and Schmidt (2010: 201) add that when conceiving of humanity as a unified whole with nature "… any adequate ethic for water, the environment or natural resource policy, must begin by situating moral claims within the relationships upon which humans depend for life."

A further approach has been the extension of the human rights framework to incorporate more-than-human things (c.f. Postel 2008). For example, there has been recent legal innovation in New Zealand, when, in March 2017, the Whanganui River was given legal aspects of "personhood" (Roy 2017). In this case, the river can be understood to act as a person in a court of law. New Zealand's legal innovation was subsequently cited in India with reference to the Ganges River which for a brief period was granted similar rights by the Uttarakhand State High Court, before being overruled three months later by the Supreme Court (BBC 2017; Safi 2017). Whilst still human-centric in a sense, probably so as to fit with the existing legal structure that itself is anthropocentric, it would suggest a shift in what has intrinsic value and, in this case, legal standing. There have also been other earlier precedents, for example, in Ecuador, in 2008, the constitution enshrined nature's "right to integral respect for its existence and for the maintenance and regeneration of its life cycles, structure, functions and evolutionary process." This positive rights approach was subsequently adapted in Bolivia (La Follette & Maser 2017). Despite this, economic pressures such as oil extraction and mining continue to degrade nature.

7.7. Situational and plural ethics

The so-called modernisation of energy, food, and water systems has led to a push towards uniformity, initially facilitated by the "command and control" function of the state (Groenfeldt 2010), and the emergence of powerful state bureaucracies (Molle, Mollinga, & Webster 2009). This shift towards modern forms of state building reflected more fundamentally the emergence of the state's ideology with respect to the control and domination of nature as a process for achieving modernisation (Coronil 1997; Mol & Sonnenfeld 2000). This push towards uniformity and standardisation was further facilitated by globalisation and market approaches (for water, see Bakker 2002; Finger & Allouche 2002; for food, see for example Goss & Burch 2001; on energy, see Jegen & Wüstenhagen 2001). The growth of powerful transnational corporations in a global free market environment with monopolistic tendencies has rendered them more powerful than many individual states, and thus highly influential in shaping standardised global value chains than simplify production and consolidate control over it (Korthals 2015).

Such an approach, associated with the imposition of universalist ethics promulgated within bureaucratic administrations and market-orientated policies, has led to the displacement of local systems of resource governance and their associated values (Bakker 2007; Swyngedouw 2005). From this imposition has often emerged a strong sense of injustice and regularly resulted in various forms of conflict (Boelens, Vos, & Perreault 2018; Lazarus et al 2011; Zwarteeven & Boelens 2014). Recognising the diversity of ethics that are situated in local places and social relations is central to understanding and working towards resolving these injustices. As Boelens, Vos, and Perreault (2018: 2) note:

> "This entails an acknowledgement of diversity and plurality – in views, knowledge, rights systems, ideas, and norms about fairness etc. – without embracing a stance of cultural relativism or denying the broader similarities across specific instances of injustice (Roth et al, 2005)"

Thus, to shift away from universalistic notions of ethics entails taking a relational and grounded perspective to render visible the values and ethics of diverse groups of people embedded in particular places, contexts, and histories (Boelens, Vos, & Perreault 2018). If the debate around the nexus approach has broadened the range of issues that need attention, it has also simultaneously deepened it by highlighting the need to give more space to a grounded "toad's eye science" to counter-balance an "eagle's eye science" that provides broad perspective but misses out critical local details (Gyawali & Thompson 2016). Such a deepening not only provides space for local perspectives and values but also brings forth new facts that are often invisible to large global agencies and corporations operating at rarefied heights. And this is what the next section on a more pluralistic and integrative social science (cultural theory) hopes to explore.

7.8. Cultural theory and a nexus ethics

Cultural theory, the neo-Durkheimian theory of plural rationalities discussed in earlier chapters, argues that many of the issues of risks and uncertainties are not only socially constructed but so constructed in a limited number of ways. In fact, using what is called Grid-Group analysis and using two discriminators along the X and Y axes,[6] it generates four permutations of bureaucratic hierarchism (strong grid-strong group) where group cohesion is strong, as is the set of pre-ascribed rules that prevent competition between constituent members; market individualism (weak grid-weak group) where there is weak group cohesion as well as pre-ascribed rules, a situation that allows for free and robust competition among its members; activist egalitarianism (weak grid-strong group) where group cohesion is strong but there are little or no pre-ascribed rules to enforce compliance which has to be done through consensus; and finally voter and consumer fatalism (strong grid-weak group) where there is little group support but strongly ascribed external prescriptions. These four social solidarities (or organising styles) not only define risks differently – hierarchism is risk managing, individualism is risk taking, egalitarianism is risk sensitising or amplifying, and fatalism is risk absorbing – but they also see nature very differently (as perverse or tolerant depending on how they are managed, benign and robust in bouncing back, fragile and ephemeral or fickle and capricious respectively).

In essence, the two discriminators essentially parallel the two fundamental questions of philosophy and consequently ethics: group is representative of the question "Who am I?" and grid of "What should I do?", the two questions we discussed earlier in the chapter as the historical starting points of much of the deeper reflections on the human condition. And what cultural theory posits is that the answers depend upon which organising style one belongs to because their value systems differ almost irreconcilably. These fundamental organising styles are found at all levels, from the village to the national, regional, and global commons. In a crisis or conflict situation, say for example a sinking ship, hierarchic morality would argue for women and children first with the captain going down with his ship as the honourable thing to do; individual behaviour is everyone for himself or herself; egalitarian ethics would demand that everyone sink or swim together; and fatalists would simply cope with whatever was offered. The first three active social solidarities (fatalism is passive in that they do not cognise or strategize but are strategized upon by the other three) have also their own type of power to wield. Hierarchism wields coercive, procedural power of rules and regulations; individualism wields persuasive power of advertising blandishments; and egalitarianism uses moral "holier than thou" power of critique, and for which they have been branded as the "ethics community" with ethics being understood in the more common sense of justice and inequality even though the other two solidarities have their own ethics that they are upholding.

Equally diverse is their response to technology that would harness some phenomenon in nature: hierarchism naturally gravitates towards large-scale, complex technologies (high dams, for instance, and large municipal piped water systems) where ranked and graded expertise within their organisations would have the final say; market individualism towards whatever flies at the particular occasion (groundwater pumping with individual pumps, solar on rooftops, bottled water); egalitarians more comfortable with "equalising" technologies (community-built and operated, small-scale brushwood dams or common water stand posts, but only if common pool goods were not able to satisfy the needs of all); and fatalists with whatever it was that the government and the private market operators offered. In the multiple definition of "what the problem is", each would be upholding its own ethics about what is the right technological choice and why the others are wrong (Schwarz & Thompson 1990).

Where cultural theory comes to the aid of the nexus approach as opposed to the IWRM approach discussed earlier is in defining this nebulous process of "integration", of who does it and how. These four diverse perspectives on what is the right thing to do cannot be integrated or melded through means that are coercive, persuasive, or even Green ethical grandstanding. Integration brought about through bureaucratic legislation may have some impact but it will be resisted or even undermined by market and egalitarian players, the former with their pursuit of efficiency and the latter with equity. And of course, the fatalist voters can always react by boycotting products or through reactive, negative voting to undermine the other solidarities if they have been sufficiently galvanised/persuaded by the other three. What cultural theory argues is that integration among these three diverse perspectives is brought about by what is called a democratic policy terrain of constructive engagement between them. And what emerges at the end of this engagement process is not a neat and elegant solution but a rather clumsy one (Verweij & Thompson 2006) where no one solidarity gets everything it wants, but in the political process of compromises, it gets more (and sometimes something newer[7]) than the nothing it would have if the process ended in an impasse.

This is also the essence of "problem feeding" of genuine interdisciplinarity we discussed earlier where one discipline begins to see the problem from the perspective of another discipline. Constructive engagement (as opposed to debilitating impasse in many environmental and social conflicts) is when there is no one single hegemon defining what the problem is and proposing its own solution: every organising style has a place and a voice at the table and is not only heard but is responded to as well by the others. What this does is to allow each to re-define the problem from the perspectives of the others, perhaps scale down individual expectations to get a consensus on the minimum that can be agreed upon. It will also allow different technological choices to have a hand in providing alternative solutions: water scarcity in a town could be handled from a new storage dam (hierarchic agency approach), supply of bottled water and individual tankers (individualist market approach), and conservation measures (egalitarian approach). It would be the ethical and plural democratic nexus of "many ten per cent solutions" as opposed to an imposed single solution of the dominant hegemon.

To summarise, ethics has emerged as the new frontier in further exploring the nexus approach: it opens up a vista hitherto invisible to the academic and practitioner communities seeking to understand the conundrums of development, both in the global North and the South. Thinking on this in the previous decades had been boxed in between the dualism of either market-led (economic) assessments or state-led (legalistic) procedural solutions. Examining ethical concerns not only allows bringing in other value concerns: it also allows it to do so more democratically by bringing all the voices to the policy table and fostering an engagement opening up space to innovative solutions.

Notes

1 Values are different from behaviour. Whilst individual behaviour may be the expression of values, behaviour may also be shaped by other incentives such as economic incentives or coercion (Dobson 2007).

2 Hinduism not being a single religion but rather a supermarket of different *mat* and *panths* (beliefs and practices) and Buddhism originally being a reform movement within Hinduism deriving much of its philosophy from the earlier Hindu *Samkhya* tradition, it is difficult to speak of singular Hindu ethics.

3 "More-than-human" refers to all life that is not human.

4 Early debates towards this relationship in the US were reflected in the conservation versus preservation debate over the fate of the Hetch Hetchy Valley between Gifford Pinchot and John Muir.

5 In a presentation made by Antonio Acedo Jr. of the World Vegetable Center, ICRISAT Campus, Hyderabad, Telangana, India at the Asian Institute of Technology (AIT) nexus

workshop in Bangkok on 22–23 January 2013: *Water Energy Food Nexus – Critical Role of Food Loss Reduction and AVRDC's Initiatives in the Mekong Region and Asia.* Available at: www.hydrology.nl/mainnews/1-latest-news/468-international-expert-workshop -on-the-water-energy-food-nexus.html.

6 Older versions of cultural theory use this term with group being the X-axis denoting weak or strong group affinity and grid being the Y-axis denoting weak or strong pre-ascribed rules. Newer versions have dropped the term Grid-Group and describe them as competition fettered or unfettered and transactions between members symmetrical or asymmetrical respectively. The basic principles of cultural theory and its applications to various conflicts and causes are well covered in Thompson et al (1990), Verweij and Thompson (2006), Thompson (2008), Thompson (2017), Schwarz and Thompson (1990), and Gyawali et al (2017).

7 Thompson (2008) discusses in detail, using cultural theory, the clumsy process that led to the re-locating and re-building of the Arsenal Football Stadium in London: the elegant solution proposed was to translocate it to far-away Rugby angering its local fans and caus-ing significant loss to the local business community. The new location next door and in the same Borough of Islington meant a win-win situation for all three social solidarities.

References

Allouche, J. (2016). The birth and spread of IWRM – A case study of global policy diffusion and translation. *Water Alternatives*, 9(3), 412–433.

Armstrong, A. (2009). Viewpoint: Further ideas towards a water ethic. *Water Alternatives*, 2(1), 138–147.

Bakker, K. (2002). From state to market? Water mercantilization in Spain. *Environment and Planning A*, 34(5), 767–790.

Bakker, K. (2005). Neoliberalizing nature? Market environmentalism in water supply in England and Wales. *Annals of the Association of American Geographers*, 95(3), 542–565.

Bakker, K. (2007). The "commons" versus the "commodity": Alter-globalization, anti-pri-vatization and the human right to water in the global south. *Antipode*, 39(3), 430–155.

BBC (2017). India's Ganges and Yamuna rivers are "not living entities". 7 July 2017, www. bbc.com/news/world-asia-india-40537701.

Boelens, R., Vos, J., & Perreault, T. (2018). Introduction: The multiple challenges and layers of water justice struggles. In R. Boelens, T. Perreault, & J. Vos (eds). *Water justice* (pp. 1–32). Cambridge, UK: Cambridge University Press.

Bollier, D. & Helfrich, S. (eds) (2012). *The wealth of the commons: A world beyond market and state*. Amherst, MA: Levellers Press.

Brom, F. W. (2000). Food, consumer concerns, and trust: food ethics for a globalizing market. *Journal of Agricultural and Environmental Ethics*, 12(2), 127–139.

Brown, P. G. & Schmidt, J. J. (eds) (2010). *Water ethics: Foundational readings for students and professionals*. Washington, DC: Island Press.

Carter, N. (2007). *The politics of the environment: Ideas, Activism, Policy*. Cambridge, UK: Cambridge University Press.

Colopy, C. (2012). *Dirty, sacred rivers: Confronting South Asia's water crisis*. Oxford and New York: Oxford University Press.

Comstock, G. (2010). Ethics and genetically modified foods. In F.-T. Gottwald, H. W. Ingen-siep, & M. Meinhardt (eds). *Food ethics* (pp. 49–66). Springer: New York, NY, 49–66.

Coronil, F. (1997). *The magical state: Nature, money, and modernity in Venezuela*. Chicago, IL: University of Chicago Press.

Dobson, A. (2007). Environmental citizenship: Towards sustainable development. *Sustainable Development*, 15(5), 276–285.

Finger, M. & Allouche, J. (2002). *Water privatisation: Transnational corporations and the re-regulation of the water industry*. London and New York: Spon Press.

Frigo, G. (2017). Energy ethics, homogenization, and hegemony: A reflection on the traditional energy paradigm. *Energy Research & Social Science*, 30, 7–17.

Goss, J. & Burch, D. (2001). From agricultural modernisation to agri-food globalisation: The waning of national development in Thailand. *Third World Quarterly*, 22, 969–986.

Groenfeldt, D. (2010). Viewpoint – The next nexus? Environmental ethics, water policies, and climate change. *Water Alternatives*, 3(3), 575–586.

Groenfeldt, D. & Schmidt, J. J. (2013). Ethics and water governance. *Ecology and Society*, 18(1), 14.

Gyawali, D., Thompson, M., & Verweij, M. (2017). *Aid, technology and development: The lessons from Nepal*. London: Routledge Earthscan.

Gyawali, D. & Thompson, M. (2016). Restoring development dharma with toad's eye science? *IDS Bulletin*, 47(2A), 170–189.

Gyawali, D. (2009). Water and conflict: Whose ethics is to prevail. In M. R. Llamas, L. Martinez-Cortina, & A. Mukherji (eds) (2009). *Water ethics*. Boca Raton, FL: CRC Press.

Heffron, R. J. & McCauley, D. (2014). Achieving sustainable supply chains through energy justice. *Applied Energy*, 123, 435–437.

Heffron, R. J. & McCauley, D. (2017). The concept of energy justice across the disciplines. *Energy Policy*, 105, 658–667.

Hefny, M.A. (2009). Water management ethics in the framework of environmental and general ethics: The case of islamic water ethics. In M. R. Llamas, L. Martinez-Cortina, & A. Mukherji (eds) (2009). *Water ethics*. Boca Raton, FL: CRC Press.

Ho, E. (2015). Bound by ethical complexities and socio-material histories: An exploration of household energy consumption in Singapore. *Energy Research & Social Science*, 10, 150–164.

Jegen, M. & Wüstenhagen, R. (2001). Modernise it, sustainabilise it! Swiss energy policy on the eve of electricity market liberalisation. *Energy Policy*, 29(1), 45–54.

Jenkins, K., McCauley, D., Heffron, R., Stephan, H., & Rehner, R. (2016). Energy justice: A conceptual review. *Energy Research & Social Science*, 11, 174–182.

Jenkins, K., Sovacool, B. K., & McCauley, D. (2018). Humanizing sociotechnical transitions through energy justice: An ethical framework for global transformative change. *Energy Policy*, 117, 66–74.

Knox, J. H. (2014). Report of the Independent Expert on the issue of human rights obligations relating to the enjoyment of a safe, clean, healthy and sustainable environment. Human Rights Council: Twenty-fifth session, Agenda item 3, Promotion and protection of all human rights, civil, political, economic, social and cultural rights, including the right to development. New York: United Nations Human Rights Council.

Korthals, M. (2015). Ethics of food production and consumption. In R. Herring (ed). *The Oxford Handbook of Food, Politics, and Society* (pp. 231–252). Oxford: Oxford University Press.

La Follette, C. & Maser, C. (2017.) *Sustainability and the rights of nature: An introduction*. Boca Raton, FL: CRC Press.

Lazarus, K., Badenoch, N., Dao, N., & Resurreccion, B. (eds) (2011). *Water rights and social justice in the Mekong region*. London: Earthscan.

Leopold, A. (1949). The land ethic. In *A Sand Country Almanac* (pp. 167–189). Oxford: Oxford University Press.

Light, A. & Rolston III, H. (eds) (2002). *Environmental ethics: An anthology*. Malden , UK: Wiley-Blackwell.

Liu, J. P., Dorjderem, A., Fu, J., Lei, X., Liu, H., Macer, D., Qiao, Q., Sun, A., Tachiyama, K., Yu, L., & Zheng, Y. (2011). *Water ethics and water resource management*. Bangkok: UNESCO Bangkok.

Llamas, M. R., Martinez-Cortina, L., & Mukherji, A. (2009). *Water ethics*. Boca Raton, FL: CRC Press.

Lopez-Gunn, E., De Stefano, L., & Llamas, M.R. (2012). The role of ethics in water and food security: balancing utilitarian and intangible values. *Water Policy*, 14, 89–105.

Malik, K. (2014). *The quest for a moral compass: A global history of ethics*. London: Atlantic Books.

Mason, K. & Milbourne, P. (2014). Constructing a "landscape justice" for windfarm development: The case of Nant Y Moch, Wales. *Geoforum*, 53, 104–115.

Mol, A., & Sonnenfeld, D. (2000). Ecological modernisation around the world: An introduction. *Environmental Politics*, 9, 3–14.

Molle, F. (2008). Nirvana concepts, storylines and policy models: Insights from the water sector. *Water Alternatives*, 1(1), 131–156.

Molle, F., Mollinga, P. P., & Wester, P. (2009). Hydraulic bureaucracies and the hydraulic mission: Flows of water, flows of power. *Water Alternatives*, 2(3), 328–349.

Moore, G. E. (1959). *Principia ethica*. Cambridge, UK: University Press.

Moss, J., McMann, M., Rae, J., Zipprich, A., Macer, D., Nyambati, A. R., Ngo, D., Cheng, M-M., Manohar, N., & Wolbring, G. (2011). *Energy equity and environmental security*. Bangkok: UNESCO Bangkok.

Nowell-Smith, P. H. (1954). *Ethics*. London: Penguin Books.

Ostrom, E., Gardner, R. and Walker, J. (1994). *Rules, Games and Common Pool Resources*. University of Michigan Press: Michigan.

Ostrom, E., Burger, J., Field, C. B., Norgaard, R. B., & Policansky, D. (1999). Revisiting the commons: Local lessons, global challenges. *Science*, 284, 278–282.

Pahl-Wostl, C., Gupta, J., & Petry, D. (2008). Governance and the global water system: a theoretical exploration. *Global Governance*, 14, 419–435.

Pinker, A. (2018). Tinkering with turbines: Ethics and energy decentralization in Scotland. *Anthropological Quarterly*, 91(2), 709–748.

Postel, S. (2008). The missing piece: A water ethic. *The American Prospect*, 19(6). Available at: www.prospect.org/cs/articles?article=the_missing_piece_a_water_ethic.

Preiser, R., Pereira, L. M., & Biggs, R. (2017). Navigating alternative framings of human-environment interactions: Variations on the theme of "Finding Nemo". *Anthropocene*, 20, 83–87.

Putnam, H. (2005). *Ethics without ontology*. Cambridge, MA: Harvard University Press.

Rai, J. S., Thorheim, C., Dorjderem, A., & Macer, D. (2010). *Universalism and ethical values for the environment*. Bangkok: UNESCO Bangkok.

Roth, D., Boelens, R., & Zwarteveen, M. (eds) (2005). *Liquid relations: Contested water rights and legal complexity*. New Brunswick, NJ: Rutgers University Press.

Roy, E. A. (2017). New Zealand river granted same legal rights as human being. *The Guardian*, 16 March 2017. www.theguardian.com/world/2017/mar/16/new-zealand-river-granted-same-legal-rights-as-human-being.

Safi, M. (2017). Ganges and Yamuna rivers granted same legal rights as human beings. *The Guardian*, 21 March 2017. www.theguardian.com/world/2017/mar/21/ganges-and-yamuna-rivers-granted-same-legal-rights-as-human-beings.

Schlör, H., Venghaus, S., Fischer, W., Märker, C., & Hake, J. F. (2018). Deliberations about a perfect storm–The meaning of justice for food energy water-nexus (FEW-Nexus). *Journal of Environmental Management*, 220, 16–29.

Schmidt, J. J. (2010). Water ethics and water management. In P. G. Brown & J. J. Schmidt (eds). *Water ethics: Foundational readings for students and professionals* (pp 3–15).Washington, DC: Island Press.

Schmidt, J. J. (2017). Social learning in the Anthropocene: Novel challenges, shadow networks, and ethical practices. *Journal of Environmental Management*, 193, 373–380.

Schwarz, M. & Thompson, M. (1990). *Divided we stand: Redefining politics, technology and social choice.* London: Harvester Wheatsheaf.

Selborne, L. (2000). *The ethics of freshwater: A survey.* Paris: UNESCO.

Singer, P. (1975). *Animal liberation: A new ethics for our treatment of animals.* New York: Random House.

Sison, A. J. G. (2009). Water and wisdom as embodied in the works of Thales of Miletus. In M. R. Llamas, L. Martinez-Cortina, & A. Mukherji (eds) (2009). *Water ethics.* Boca Raton, FL: CRC Press.

Sovacool, B. K. & Dworkin, M. H. (2015). Energy justice: Conceptual insights and practical applications. *Applied Energy*, 142, 435–444.

Sovacool, B. K., Burke, M., Baker, L., Kotikalapudi, C. K., & Wlokas, H. (2017). New frontiers and conceptual frameworks for energy justice. *Energy Policy*, 105, 677–691.

Suhardiman, D., Lebel, L., Nicol, A., & Wong, T. (2017). Power and politics in water governance: Revisiting the role of collective action in the commons. In D. Suhardiman, A. Nicol, & E. Mapedza (eds). *Water governance and collective action: Multi-scale challenges* (pp 9–20). London: Earthscan.

Swyngedouw, E. (2005). Dispossessing H2O: The contested terrain of water privatization. *Capitalism, Nature, Socialism*, 16, 1–18.

Thompson, M. (2008). *Organising and disorganising: A dynamic and non-linear theory of institutional emergence and its implications.* Axminster, UK: Triarchy Press.

Thompson, M. (2017). *Rubbish theory: Creation and destruction of value.* London: Pluto Press.

Thompson, M., Ellis, R., & Wildavsky, A. (1990). *Cultural theory.* Boulder, CO: Westview Press.

Verweij, M. & Thompson, M. (eds) (2006). *Clumsy solutions for a complex world: Governance, politics and plural perceptions.* London: Palgrave Macmillan.

Warnock, M. (1998). *An intelligent person's guide to ethics.* London: Duckworth Overlook.

Zwart, H. (2000). A short history of food ethics. *Journal of Agricultural and Environmental Ethics*, 12(2), 113–126.

Zwarteveen, M. Z. & Boelens, R. (2014). Defining, researching and struggling for water justice: Some conceptual building blocks for research and action . *Water International*, 39(2), 143–158.

8

CONCLUSION

"Democratising" the nexus

The nexus as a concept is a compelling idea. This book has argued that the nexus needs to be reconceptualised, however, away from a technical solution to natural resource scarcity, which is apolitical in origin and intent, and towards a clear and articulated political choice about allocation and trade-offs between resources and the imagining of the future of water–food–energy systems and their interlinkages. The politics of the nexus need to be understood along three axis, politics of knowledge, politics of indifference, and international political economy and geopolitics. Current nexus policies reflect particular framings around technology, mass production, and water/energy/food securities, as well as specific environmental economic models emphasising efficiency and trade-offs. These framings are driven by food and energy needs that are centred upon a modernisation-biased urban-centric system, where production unquestionably is to serve these needs rather than also taking into account the consequences of these needs environmentally and socially.

In this book, we have traced the concept of nexus produced and circulated within research agendas and global policy circles. We show how the origins of the nexus in its current incarnation(s) is rooted in the global food and energy price shocks of 2007/8 from which emerged narratives of water–food–energy scarcity and global crisis, echoing earlier "limits to growth" arguments. These alarmist narratives, we have argued, are highly political statements that have led to a nexus-agenda that draws heavily on market-technical framings in terms of research, analysis, and proposed policy solutions, for example promoting "sustainable intensification" and other green economy strategies for resource management that are often about increasing productivity through resource efficiency, widening the extent of market-solutions and resource commodification, and furthering state–private sector collaborations. Such approaches largely affirm the current global production-consumption systems in place, proposing measured reforms, and downplay alternatives both locally and at higher scales that propose more profound transformation. They do not seriously problematise or engage with the need to address distributive inequalities, resource grabbing, and enclosure of commons, or systemic sources of social injustice at scales ranging from the local to the global. In its current formulations, we doubt whether the market-technical rendering of the nexus can lead to sustainable and just outcomes.

Whether the nexus is anything new is also an important question. In this book, we have examined the continuities and discontinuities from the nexus' predecessor, namely Integrated Water Resources Management (IWRM), which is also based on the idea that there should be cross-sectoral integration to accommodate the need for water for people, water for food, water for nature, and water for industry. We suggest that basic principles of integration remain the same (policy coherence, building institutional connections, etc), whilst there are also key differences such as scale – whilst IWRM emphasises river basin scales, nexus opts for a wider range of system scales (with the boundaries still much disputed). Furthermore, whilst nexus in principle was intended to shift away from a water-centric focus and to place water, energy, and food on more equal footings, in practice this has not been achieved with water security in many cases still remaining a key organising principle. More fundamentally, however, we flag a concern in the past raised towards IWRM, namely inclusiveness of decision-making and the consequences of rendering technical, as an unfortunate continuity from IWRM to the nexus.

Moving beyond the current global nexus discourses, we have sought to draw attention to the interaction between food, water, and energy concerns in formal and informal institutions within hybrid governance systems, emphasising the significance of local and national history and practices. Drawing in particular on case studies from Nepal, we draw out how systems have become both de-nexused and have the potential to be nexused again. Our main argument here is that there is a need to move beyond technocratic/bureaucratic integrated management approaches and towards pluralised policy terrains at a more basic social level that emphasise bottom-up approaches and that contain a range of styles of social organisation, namely hierarchism, individualism, and egalitarianism. This approach accommodates *inter alia* a greater diversity of voices, interests, values, rationalities, strategies, technologies, and forms of knowledge.

Synthesizing the book's key findings, we put forward four key concluding points.

Firstly, the nexus is a contested discourse. Discourses – as relations of discursive practice – proscribe what should be considered as legitimate knowledge and perspectives, and therefore also what knowledge and perspectives are excluded. Two dominant distinct nexus discourses have emerged, even as we note that in many ways nexus-thinking is not new: nexus as a discourse promoting resource optimisation for economic efficiency, as originating and promoted by the World Economic Forum since 2008; and nexus as a discourse for promoting sustainable development, as promoted at the Rio+20 conference in 2012. These are placed in relation to more critical thinking on the nexus as an idea amongst some academics and civil society groups. The analysis of these nexus discourses, the organisations that have promoted them, and the epistemes they are seeking to create reveal the power inequalities within them and thus whose interests are privileged or downplayed, and how they are shaping food, energy, and water systems, and their governance.

Secondly, the integrative imaginary of these three systems into one system is unclear in terms of entry point. Seeing the system from a water perspective and its relationship with food and energy is very different than seeing it from an energy or a food perspective. In fact, these three systems are part of different international political economic systems in terms of production, consumption, and waste. In contrast to a prescribed optimal scientific rationale system, one can observe two sub nexuses, water–food trade on the one hand and energy–climate change on the other. This book has also highlighted how the idea of policy integration has proven to be difficult to implement in the past. Water–food–energy nexus framing and policies neglect the very important role of institutionally mediated human agency and the resistance to change. Although the coupling of water, food, and energy use has increased over time and previously independent institutions need to adapt to take into account cross-sector linkages, vested interests and norms may prevent such adaptation and reflect institutional path dependencies.

Thirdly, the issue of scale and power and top-down vision may not take into account lived experiences of the nexus from a different perspective. As exemplified through the case study of water storage and hydropower in Nepal in Chapter 5, which lies at the intersection of water, food, energy, and climate change concerns, a nexus approach emerges without having to wait for a statesman messiah or a major disaster. Thus storage and transport interstices under considerations of efficiency and footprints are areas where market individualism has to bring forth technological innovations, bureaucratic hierarchism has to bring regulatory innovations, and activist egalitarianism has to bring about innovations in behaviour and value changes. There is a need to ask how the nexus policy terrain can be pluralised, how space can thereby be provided to multiple voices who define very differently what the problem is.

Fourthly, the interests and current framing of the nexus does not take into account the policy priorities of small scale users and the most marginalised. This book aimed at rediscovering these local nexuses, these set of de-nexused activities where more research is needed on how the nexus can be grounded and conceptualised from below, from the perspective of the farmer, the fisher, and so on. While decisions on the water, energy, and food may be nexused at the level of the farming household, the process of silo-fication gains prominence and strength as once moves to higher levels of governance. Global priorities need to better connect with local concerns and the politics of scale should be made more explicit and acknowledged in nexus debates. More fundamentally, the politics of knowledge and expertise need to recognised, leading to a "reflexive turn" that treats the framing of nexus problems and solutions as a matter of political contestation.

This last point brings the challenge of transdisciplinarity to the fore. The integration of science with society including local knowledge and value systems to solve real world problems where uncertainties are high and the underlying values are being challenged. As a concept, the nexus is supported by a rapidly growing evidence base and a community of practitioners and policy makers, providing a powerful but largely disconnected knowledge base to understand the relationships and trade-offs between the different sectors and disciplines characterising the nexus. What is needed, is not only "joined up thinking", but profoundly transformative change in infrastructures, organisations, behaviours, markets, governance practices, and even cultures more widely. Transdisciplinary nexus research should aim to understand the world not as a messy realm of competing value systems from which research should abstract to arrive at transcendental and disinterested truths but as an inevitable condition they have to appreciate and learn from. Research practices should offer spaces for friction by allowing competent colleagues and non-academics to object and induce other modes of thinking. The interdisciplinary and multi-sited approach suggested in this book allows us to conceive of more multidimensional understandings of the politics of the nexus. These include political economy and political ecology (with an accent on material and structural forms of power and their implications for questions of access and justice), to institutional politics (focusing on national and global organisational forms), to discursive expressions of power (through knowledge and values).

Overall, technological and managerial solutions are put forward without taking into account the current inequalities in the system. A clear normative positioning on the nexus around equity, social progress, and environmental justice is needed. The consequences of any nexus policy will have distributional consequences. Developing a nexus ethics is something that comes into its own at the point just before action is about to commence. Key questions to ask are: Is an action just? Correct? Necessary? Ennobling? Different ideas of nexus ethics are based on different perspective linked to cultural theory where each way of organising – market individualism, bureaucratic hierarchism, and civic egalitarianism – is based on its own rationality that upholds the ethics that defines who it is, what it believes in, and what it should do about it.

As the nexus policy framing process unfolds, many are questioning whether, again, this will prove to be an ineffective discourse that drives only more rhetoric, bureaucracy, and managerialism. All these points are not just limited to the nexus debate but echo some of the key debates around green economy and transformations, planetary boundaries and the Anthropocene/Capitolocene, and the dominance of "technocentric", top-down, and marketised view of sustainability and transformation. They all reveal how clear urgencies and imperatives restrict the contours of legitimate political debate and instead how these problems need to be opened up for inclusion, deliberation, democracy, and justice. Synergies and win-win constructed nexus policy framings often obscure the many hard trade-offs implied by attempts to square environmental aims with social justice. Rather than there being one big nexus, it is more likely that there will be multiple nexused and de-nexused sets of framing and ideas that will intersect, overlap, and conflict in unpredictable ways. And this – a properly nexused world – is precisely where a rethinking of democracy and its practice will be critical.

INDEX

Page numbers in *italics* refer to figures. Page numbers in **bold** refer to table. Page numbers followed by "n" refer to notes.

For Product Safety Concerns and Information please contact our EU
representative GPSR@taylorandfrancis.com
Taylor & Francis Verlag GmbH, Kaufingerstraße 24, 80331 München, Germany

www.ingramcontent.com/pod-product-compliance
Lightning Source LLC
Chambersburg PA
CBHW050530270326
41926CB00015B/3157